More Clouds, More Warmth? Understanding Cloud Feedbacks in the Arctic

Raviie

Contents

Chapter 1.

Introduction

1.1. Arctic Amplification

The anthropogenic emission of greenhouse gases, mainly carbon dioxide (CO_2), has led to globally increasing temperatures. The Arctic (defined as north of $60°$ N unless stated otherwise) climate is warming faster than the rest of the planet. This effect is often called "Arctic Amplification" (AA) and has been observed and predicted by climate models. It is also reproducible for previous natural climate changes (Miller et al. 2010). With this current warming, the Arctic system is changing in many different ways. There are changes in atmospheric transport pathways (Mewes et al. 2019) of energy, moisture, and aerosol, as well as local changes in fluxes of energy and moisture caused by a decline in snow cover on land, a decline in sea-ice concentration, and changes in cloud properties. These changes can have positively reinforcing effects (so called feedback mechanisms) and are the reason for the AA (Manabe et al. 1980; Miller et al. 2010; Overland et al. 2011).

One such feedback mechanism is the surface albedo feedback. It is caused by melting of snow and ice following an increase in near-surface temperatures, which lowers the albedo of the surface, leading to a higher absorption of solar radiation, heating the surface further.

The water vapour feedback is caused by an increase in the surface temperature, which leads to higher water vapour concentrations through enhanced evaporation. This water vapour absorbs thermal radiation coming from the Earth's surface, further increasing the near surface temperatures.

The lapse-rate feedback is caused by varying degrees of warming throughout the vertical extent of the atmosphere. In the Tropics strong convection in clouds causes a coupling between the surface temperature and that in the free troposphere, and a near moist adiabatic temperature profile. In a warmer, more moist climate more latent heat is released aloft, further steepening the tropical temperature profile, causing a negative lapse rate feedback (Bony et al. 2006). There a smaller surface warming is required to offset a TOA radiative imbalance (Pithan et al. 2014). In the Arctic, the warming is largely confined to the lower level of the atmosphere because of the stable planetary boundary layer typical for this region. Thus the opposite of the tropics is true. The lapse rate feedback is positive in the Arctic (Pithan et al. 2014; Block et al. 2020).

Arctic clouds have a net warming effect on the Arctic, because of a combination of the high surface albedo regularly exceeding 80 % (Istomina et al. 2015), and low solar zenith angles. When clouds are located over a high surface albedo, they do not increase the planetary albedo, which is a cooling effect in other regions. The low solar zenith angles can lead to clouds reflecting light downward that was reflected by a bright surface before. The Arctic clouds are expected to increase in occurrence in a warmer Arctic, leading to a positive feedback. A major reason for expecting a more cloudy Arctic is increased evaporation from a warmer and more open ocean (e.g. Vavrus et al. 2011). The net effect of clouds in a warming Arctic is however uncertain, since the capability of clouds to increase the planetary albedo becomes more important with the declining Arctic surface albedo caused by snow and ice melt. This constitutes a negative feedback (Goosse et al. 2018).

As near surface temperatures rise in the Arctic, the outgoing thermal radiation increases which constitutes a loss of energy for the system. This is called the Planck feedback. However, because the outgoing energy increases with the fourth power of the temperature, this loss is smaller at low temperatures. To achieve the same increase in upward thermal energy flux the surface temperatures have to increase more than in warmer regions.

Aerosols of natural and anthropogenic origin can influence these feedback mechanisms by interactions with the radiative fluxes in the atmosphere, changing the albedo of snow and ice, or by impacting cloud formation and lifetime. Therefore they have the potential to impact the AA considerably. Black carbon (BC) is of special interest since it is strongly absorbing in the visible light range of the electromagnetic spectrum, which makes it especially important for aerosol radiation interactions.

The strength and relative importance of these feedbacks remains uncertain partly because they are interlinked (Wendisch et al. 2017). The German collaborative research centre "(AC)^3" focusses on these feedbacks to further the understanding of this rapid climate change in the Arctic.

1.2. Arctic aerosol

Compared to other regions, the Arctic air is relatively clean in terms of aerosol particles with annual median number concentrations (e.g. at ZOTTO station, Siberia, 60.79° N 89.35° E) of 570 cm^{-3} (Heintzenberg et al. 2011) compared to the German rural background (Melpitz, 51.53° N 12.93° E) with 7400 cm^{-3} (Sun et al. 2019). The annual median BC mass concentrations are 30 ng m^{-3} (Eleftheriadis et al. 2009) for the Arctic (Ny-Ålesund, 78.90° N 11.83° E) and 500 ng m^{-3} (Sun et al. 2019) for Germany (Melpitz).

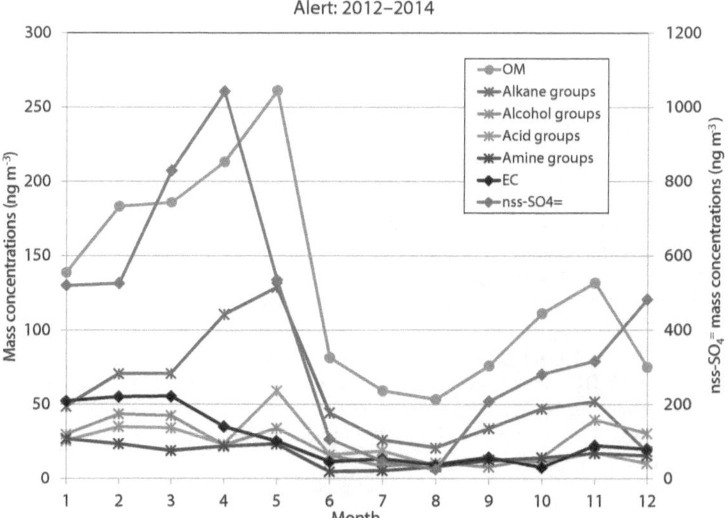

Figure 1.1.: Arctic aerosol chemical composition as observed in Alert (82.45° N 62.51° W) by Leaitch et al. (2018). Monthly averages for the years 2012 to 2014 of organic mass (OM), elemental carbon (EC), and non-sea-salt SO_4 (nss-SO4=).

The Arctic aerosol is mainly composed of organic carbon (OC), sulphate, mineral dust, sea salt, and BC. Figure 1.1 shows the chemical composition of Arctic aerosol as observed on average for the years 2012 to 2014 in Alert (82.45° N 62.51° W, Leaitch et al. 2018). While sulphate, mineral dust, and OC have by far the largest contribution in terms of aerosol mass, BC is interesting because of its aforementioned optical properties. Sea salt also contributes to the Arctic aerosol load and is mainly found in low altitudes. With the cold temperature, special sunlight conditions during the polar day and polar night, snow and ice covered surfaces, and related weather phenomena, like persistent low level clouds and a stable stratification of the lower troposphere, the conditions for the Arctic aerosol are unique.

1.2.1. Seasonality

The aerosol concentration in the Arctic shows a strong seasonality as seen, e.g. in the vertical profiles of the BC mass mixing ratios in Figure 1.2. The profiles were simulated for the year 2001 with the Canadian global air quality model GEM-AQ (L. Huang et al. 2010). The highest concentrations near the surface are reached in winter and early spring (in black), a phenomenon often referred to as "Arctic Haze". It describes a period of high aerosol optical thickness (AOT), during which not only the maximum concentrations of BC, but also that of sulphate are found within the lowermost part of the atmosphere. This is caused by a low deposition efficiency, and an efficient transport in the winter and spring months (Shaw 1995; Stohl 2006; Quinn et al. 2007), the cause of which is discussed later.

AOT is commonly used as a measure for the vertically integrated aerosol extinction. This can be seen in Figure 1.3, where higher AOT can be found for spring (Breider et al. 2014), both in the model GEOS-Chem, and observations by the AERONET network (Holben et al. 1998). The time of this maximum varies with the exact location between late winter and early spring (Quinn et al. 2011). In summer the near surface BC concentrations typically are at their minimum, because of a considerably higher wet deposition rate, combined with a lower northward transport, as also discussed in more detail later. At higher altitudes, the summer BC concentrations can be higher than in winter, as also shown in Figure 1.2.

The aerosol particles are not horizontally evenly distributed throughout the Arctic region. This can also be seen in Figure 1.3, indicated by the AOT. The AOT, and therefore the total aerosol burden, are on average higher in the Eurasian Arctic than in the American Arctic. This is caused by stronger aerosol sources in the eastern hemisphere, both within the Arctic and also further south.

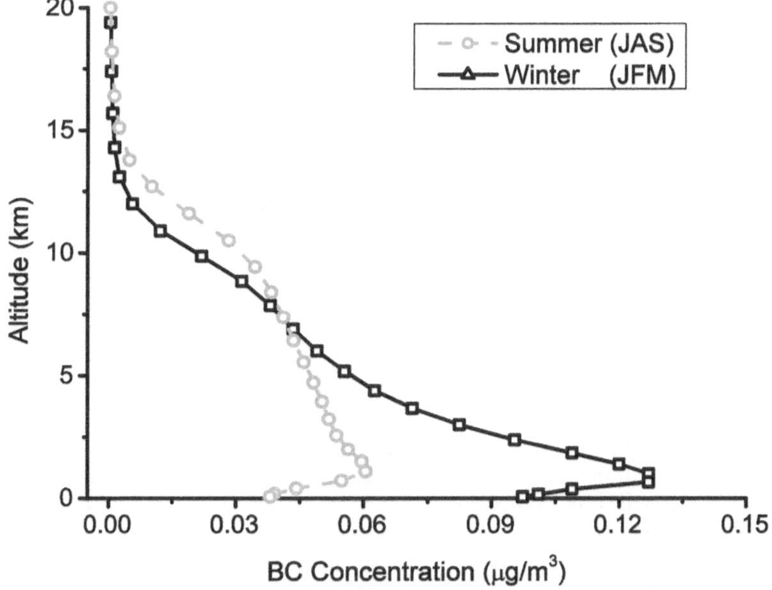

Figure 1.2.: Arctic (70°-90° N) vertical BC profiles as simulated by L. Huang et al. (2010). Seasonal averages of summer (July-August-September) and winter (January-February-March) for the year 2001.

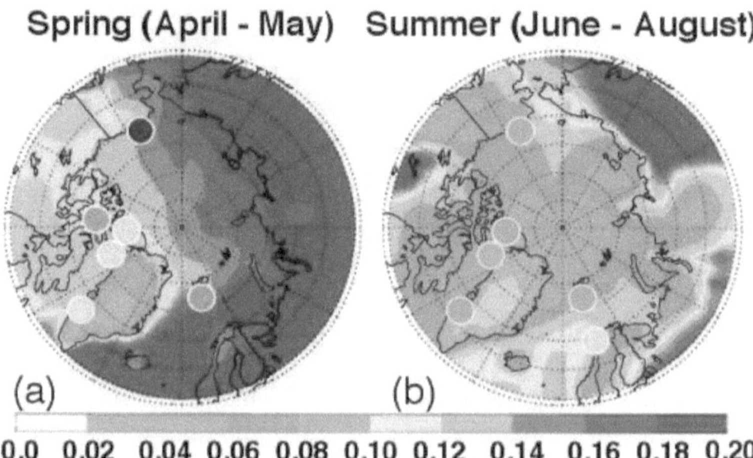

Figure 1.3.: Arctic aerosol distribution indicated by aerosol optical thickness (AOT) for spring (a) and summer (b) of 2008, as modelled by GEOS-Chem (map) and observed at AERONET stations (Holben et al. 1998) north of 65° N (circles). Taken from Breider et al. (2014).

1.2.2. Sources and transport

Aerosol emissions can occur within the Arctic from natural as well as anthropogenic sources. One natural source is the Arctic Ocean, emitting sea salt as well as OC and biogenic aerosol precursors. The biogenic emissions from the Arctic Ocean are expected to increase with declining sea-ice cover (Gilgen et al. 2018). High latitude fires emit OC and BC. Arctic fires have been especially severe in 2019 and 2020, as can be seen in the fire radiative power made available by the Global Fire Assimilation System (GFAS, Kaiser et al. 2012). High latitude dust sources such as retreating glaciers also become more important (Groot Zwaaftink et al. 2016).

The anthropogenic sources include emissions from industry and settlements. Shipping traffic is another anthropogenic source, which is expected to increase with an increasingly sea-ice free Arctic (Gilgen et al. 2018).

Most of the Arctic's aerosol particles do not originate within the Arctic. They are transported over long distances, but the dominant source regions differ between the subregions within the Arctic (Willis et al. 2018). During this transport, the composition of the aerosol particles changes. They, for example, become internally mixed with other substances, coated, or oxidised. Since local sources are relatively small, aerosol is typically more aged on average in the Arctic than elsewhere, especially during the season of the Arctic Haze.

Most of the time, the transport of aerosol to the Arctic is strongly limited by the thermodynamic condition of the air preventing a mixing of the Arctic and mid-latitudinal air. At the polar dome, the isentropes (surfaces of constant potential temperature) tend to not cross into the Arctic in the lower troposphere, but are instead forming a dome like structure, which suppresses the exchange of air. The reason is that an adiabatic lifting occurs along these surfaces, instead of a mixing with the denser cold air closer to the earth's surface (Barrie 1986; Stohl 2006). Isentropic mixing can occur in the presence of cloud formation, radiative heating or turbulence, all of which are not, or a minor factor in Arctic winter. This barrier of air masses is also often referred to as the Arctic front (Quinn et al. 2015). Its position at the surface is sketched in Figure 1.4 as blue and orange lines, for winter and summer, respectively. The dome structure of lines of same potential temperatures is indicated by the dashed lines for summer only, but is not to scale in its vertical extent.

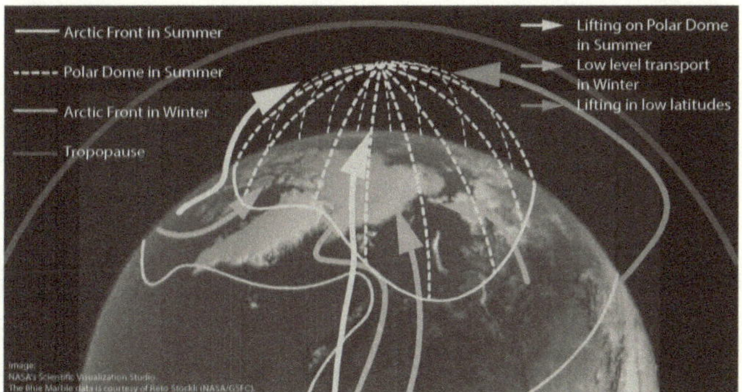

Figure 1.4.: Schematic view of the transport pathways to the Arctic (after Quinn et al. 2015). The different transport pathways are indicated by arrows (Stohl 2006). The mean position of the Arctic front for summer and winter is indicated by the orange and blue lines, respectively (Li et al. 1993).

Aerosol particles reach the Arctic through three main transport pathways (Stohl 2006). They are sketched in Figure 1.4.

1) **Low level transport:** A low altitude transport from the mid latitudes, which is mainly active in wintertime. It is only possible when the mid-latitudinal air has cooled enough for the Arctic front to extend far enough south to reach the emission sources of Eurasia (blue line in Figure 1.4). This situation is created by a similar surface cooling as in the Arctic, both caused by low solar energy influx and snow-covered surfaces. In this situation, a pollution transport through the lower Arctic troposphere is possible in a timespan of 10 to 15 days (Stohl 2006). This transport pathway is depicted by blue arrows in Figure 1.4. The top of the Greenland ice sheet is not affected by this low level transport.

2) **Lifting on polar dome:** In the summer, there are periods of northward transport

of aerosol at low altitudes. Once it reaches the Arctic front, this is followed by a lifting before being transported further north above the polar dome.

3) **Lifting in low latitudes**: The aerosol particles can also be lifted at the low latitudinal source regions for example by convection. After this, a transport northwards in the upper troposphere into the polar region is possible.

Of these three main pathways, 2) and 3) require a slow descent of the particles by radiative cooling in the Arctic, or a mixing into the polar dome. The fast low level transport 1) in combination with an inefficient removal of aerosol in winter cause the Arctic Haze (Shaw 1995).

1.2.3. Deposition

The efficiency of the transport depends on the removal of the aerosol. It is typically divided into dry deposition and wet deposition. Wet deposition, the scavenging of aerosol particles by hydrometeors, is the more important of the two deposition pathways. By using the global aerosol climate model ECHAM-HAM, it is estimated that more than 85 % of aerosol particles are deposited through wet deposition (Croft et al. 2010).

Especially the pathway of remote lifting 3) is strongly affected by the efficiency of wet deposition in the area of lifting. This is the case because the required lifting is often directly related to strong convective precipitation, causing a wet deposition of aerosol through in-cloud and below-cloud scavenging. The remotely deposited aerosol will then never reach the Arctic. The smaller fraction of aerosol that does not get removed in this way will stay at high altitudes within the Arctic for a long time. The observed typical size for background BC (not within a biomass burning plume) in the lower most stratosphere is around 100 nm with a coating thickness of either around 50 nm or around 120 nm, depending on the source type and co-emitted species (Ditas et al. 2018). The Arctic clouds in the of this aerosol tend to be pure ice clouds, which produce very little precipitation. The lifetime within the Arctic of aerosol transported by this pathway is therefore expected to be especially high, exceeding weeks, since isentropic mixing is required. This downward mixing is caused by radiative cooling, which occurs at a rate of only about $1\,\mathrm{K\,d^{-1}}$ (Quinn et al. 2011).

The lifting of air at the polar dome (pathway 2) can also cause cloud formation and precipitation and therefore wet deposition. Since the polar dome is located relatively far north in summer, this is within the Arctic.

The low altitude aerosol of the Arctic Haze that gets transported through pathway 1) also has high atmospheric lifetimes, because precipitation is rare during winter. Most of it is deposited by wet deposition, once precipitation becomes more common in spring. The aerosol deposited within the Arctic can, however, still have an effect on the Arctic climate.

1.2.4. Radiative effects

Arctic aerosol modulates the Arctic radiative energy balance through direct interaction with the solar and terrestrial radiation and by changing the cloud properties (Garrett et al. 2004) as well as atmospheric dynamics (Doherty et al. 2013). It therefore interacts with the feedback mechanisms that lead to the AA. The effects of the aerosol radiation interaction depend on the altitude of the aerosol particles, the underlying surface, the time of year, and the aerosol composition. BC is of special interest, since it is the aerosol species that absorbs solar radiation most efficiently. Mineral dust is important because of its effect on clouds globally, but also in the Arctic, with vertically integrated dust burdens in the range of $10\,\mathrm{mg\,m^{-2}}$, the majority of which is transported from outside of the Arctic, with 38 % being transported from Asia and 32 % from Africa (Groot Zwaaftink et al. 2016). There its interaction with solar radiation it is even more important because of the high surface albedo, which can decrease by the deposition of mineral dust (Groot Zwaaftink et al. 2016). Sulphate is typically associated with the reflection of solar radiation, an effect that is less relevant in the Arctic than elsewhere because of the high surface albedo, and is not active during polar night.

The effects of aerosol on the radiative energy balance throughout the year are sketched in Figure 1.5. In polar winter, when there is no sunlight, aerosol interacts with the radiation mostly indirectly through changing cloud properties. The strongly aged aerosol particles cause the formation of optically thicker clouds or additional clouds, which reduce the outgoing terrestrial radiation, and therefore warm the surface.

In spring, snow and ice still cover large areas of the Arctic sea and land. With increasing incoming solar radiation and decreasing snow and ice cover, the indirect radiative effect on the solar radiation becomes more important. It is caused by the aerosol acting as cloud condensation nuclei, leading to more, smaller cloud droplets, effectively increasing the combined surface of the droplets within a given cloud and in turn the albedo of the clouds (Twomey 1977).

In summer, the aerosol scatters or absorbs the solar radiation, reducing the amount that reaches the surface. In the case of BC, it absorbs the solar radiation, which warms it and the air surrounding it. This increases the downward terrestrial radiation. The high surface albedo, caused by snow and ice that persists throughout summer in some regions, leads to a more efficient absorption of solar radiation within layers of absorbing aerosol particles above (Quinn et al. 2008).

BC that is deposited on the ice and snow surfaces lowers their albedo. Therefore, more solar radiation is absorbed at the surface, which warms the surface. This temperature increase leads to a melting, which enriches the BC concentration in the topmost snow layer, because the melt water run-off leaves large parts of the deposited BC behind (Doherty et al. 2013). This causes a self enforcing feedback, because it leads to a stronger absorption and as a consequence to further heating and melting. All of these effects depend on the amount and composition of the aerosol and, therefore, on their

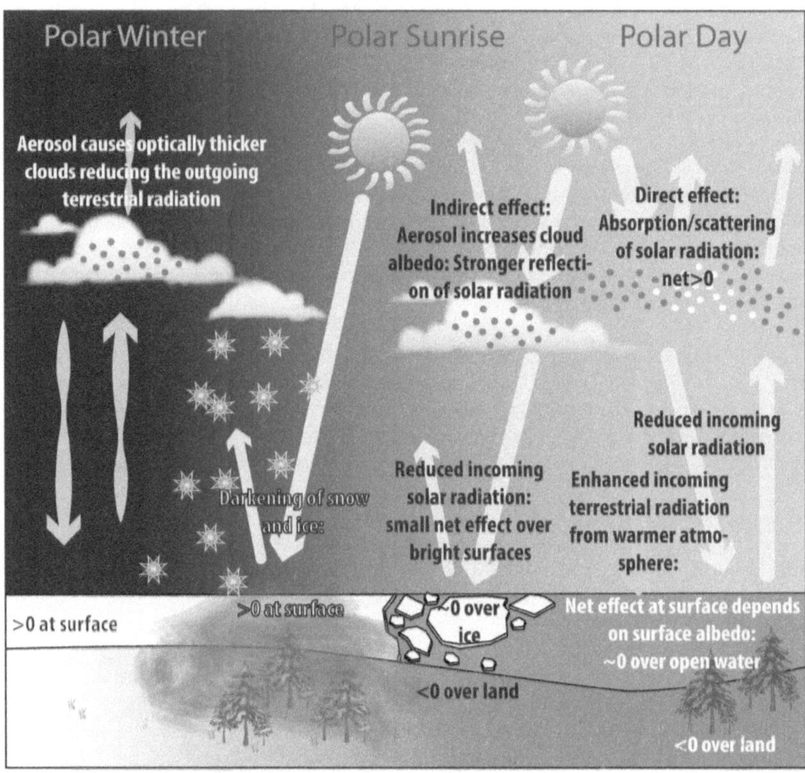

Figure 1.5.: Schematic of the way Arctic aerosol interacts with radiation. Radiation fluxes in the solar wavelength spectrum are depicted in yellow, fluxes in the terrestrial wavelength range in orange. Numbers indicate positive, negative, or neutral effects on the radiative budget in the Arctic. After Quinn et al. (2015).

sources, transport, and ageing.

1.3. Aerosol in global models

Global aerosol climate models are particularly well suited to research the effects of aerosol on the Arctic climate, because they include the various source regions and the transport pathways that are relevant for the Arctic. They also offer the opportunity to test the sensitivity of the Arctic to many processes governing the effectiveness of the transport. However, the system is very complex and simulating the life cycle of aerosol and its various effects on the global climate is challenging. In the past, models have struggled to reproduce the Arctic Haze phenomenon (e.g., Sharma et al. 2006; Quinn et al. 2007)). The problem is well illustrated in Figure 1.6, which shows the zonal mean aerosol burden of BC and sulphate for different global aerosol climate models for March and June 2008. Especially for March, but also for June, the inter-model spread in aerosol burden increases greatly from the mid latitudes to the Arctic region (Eckhardt et al. 2015).

However, other recent studies, that use more recent versions of the same or similar models, have shown great improvement in capturing the seasonality in Arctic aerosol (Vignati et al. 2010; Liu et al. 2011; Browse et al. 2012; Quinn et al. 2015). The vertical transport and seasonality of Arctic aerosol are often identified as key uncertainties (Samset et al. 2014; Arnold et al. 2016). A correct representation of the vertical aerosol distribution, however, is a prerequisite for estimating the aerosol radiative impact (Samset et al. 2013). The uncertainty in emissions and the representation of wet removal are identified as possible sources of these uncertainties (Stohl et al. 2013; Arnold et al. 2016; Winiger et al. 2017; Watson-Parris et al. 2019). To evaluate the modelled distributions and seasonality, in-situ measurements are commonly used.

Independent of that global models also differ in the representation of aerosol. Current models use modal, sectional, bulk aerosol microphysics approaches and in some cases a combination thereof (Myhre et al. 2013b). The included species also differ between models. BC and sulphate are very commonly included because of their respective important radiative properties. Sulphate, additionally to this, is the most important (in some cases only) ageing agent in global models, an important process leading to the wet removal of aerosol, as discussed in detail later. However, not all models account for an ageing of BC. All models in Phase II of the aerosol module evaluation effort AeroCom consider BC and sulphate (Myhre et al. 2013b).

At high altitudes of remote regions the AeroCom Phase II models show a tendency to overestimate BC concentrations (Samset et al. 2014). This overestimation leads to an overestimation of the anthropogenic direct radiative effect (DRE) of BC by about 25 %, compared to when the models are scaled to measurements.

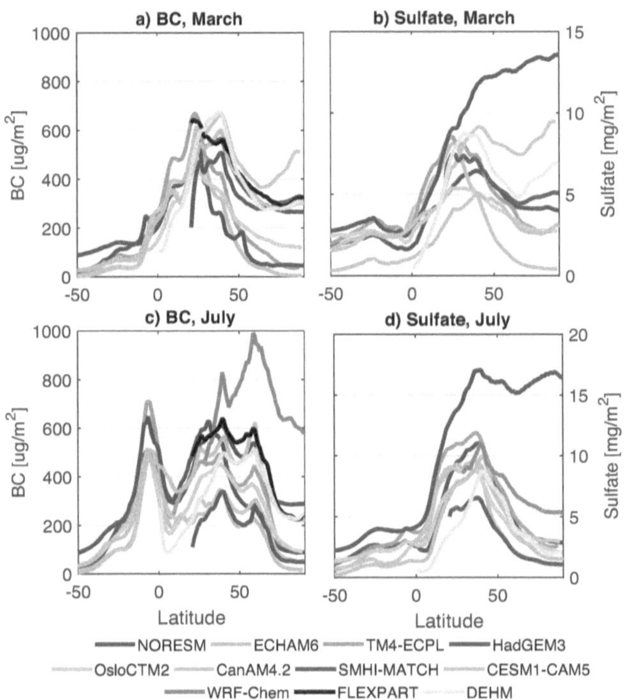

Figure 1.6.: Zonal means of BC (a, c) and sulphate (b, d) burden as simulated by different models. The top row shows monthly averages for March 2008 and the bottom row for July of the same year. After Eckhardt et al. (2015).

1.3.1. Model emissions

For anthropogenic aerosol and aerosol precursor gases, global models rely on emission datasets, while for aerosols of natural origin (e.g. sea salt and dust) emission parametrisations that depend on wind speed and other factors are commonly used. In case of anthropogenic sources and in some cases agricultural burning, the emission datasets are based on information about fuel consumption given by agencies of the emitting states and emission factors for different sectors (e.g. Bond et al. 2004; Klimont et al. 2017). Biomass burning (grass, bush, and forest fire) emissions can be derived from the burned area, the vegetation formerly located in this area, and the meteorology (Werf et al. 2017), or from the fire radiative power (Kaiser et al. 2012). Both approaches utilise measurements by satellite instruments (Justice et al. 2002).

The emissions of BC (and OC) are the species with the highest emission uncertainty, especially for open-burning (Bond et al. 2013). These emission datasets are considered to be one of the main sources of model uncertainty, second only to wet deposition (Q. Wang et al. 2014; K. Huang et al. 2015). The uncertainty range in emissions is often estimated to be within the range of about a factor of 2–3 for energy-related emissions and even higher for open-burning (e.g. Bond et al. 2007; Bond et al. 2013; Klimont et al. 2017).

1.3.2. Aerosol ageing and removal

Together with the aerosol sources, represented by the model emissions, the aerosol sinks govern the aerosol life cycle. A realistic representation of the sinks via aerosol ageing and subsequent removal is a prerequisite for realistic aerosol distributions in the model. To achieve this models use schemes of different complexity.

The deposition processes of wet and dry deposition (as mentioned in section 1.2) are typically separated in state of the art models. The dry deposition is mostly important for very large particles and is parametrised by estimating the air resistance in relation to the surface properties (Wesely 1989; Ganzeveld et al. 1998). The wet deposition is calculated as a sum of different processes, for which a multitude of parametrisations is available (Myhre et al. 2013b). Commonly considered processes are an in-cloud scavenging and a below-cloud impaction scavenging. They are typically different schemes for convective clouds and for large scale clouds and the schemes additionally discriminate between the cloud phase.

The in-cloud scavenging by liquid clouds, where aerosol acts as a cloud condensation nucleus (CCN), requires the aerosol to be hydrophilic. Since the wet deposition is a major sink for most aerosol (including BC), the ageing of aerosol from hydrophobic to hydrophilic is very important. Some models allow aerosol to age, changeing their state from hydrophobic to hydrophilic. In some models this is parametrised by an e-folding time, in others by mixing with sulphate, and in a few by a range of additional ageing agents, see Table 1.1 for some examples. One example of a very complex ageing parametrisation is one of forming layers of H_2SO_4, HNO_3, NH_3 or secondary organic aerosol on BC particles by defining a required mass fraction (He et al. 2016). Other schemes include ozone ageing of BC particles (Friebel et al. 2019). Table 1.1 also gives an overview of the type of microphysics schemes used. Depending on the type of scheme used, these scavenging processes are size dependent. Exceptions to these detailed representations are made for very long climate studies that span up to multiple millennia and use simple e-folding techniques instead to reduce computational costs.

The model error in terms of BC concentrations in the remote troposphere is considerably larger than can be explained by the uncertainty in emissions. Instead, the uncertainty in aerosol removal is the main cause for model error in the remote troposphere (Schwarz et al. 2010a; Kipling et al. 2013; Q. Wang et al. 2014). This is consistent with the finding that to reproduce BC concentration measurements in remote ocean regions a lifetime of anthropogenic BC of less than 5 days is crucial (Bauer et al. 2013; Q. Wang et al. 2014; Samset et al. 2014). This lower lifetime requires an increase in the BC wet scavenging efficiency (Jacobson 2012; Kipling et al. 2013; Q. Wang et al. 2014). However, a reduction in BC lifetime as suggested by Samset et al. (2014) is not uniformly accepted as the correct approach for improving seasonality and vertical profiles of BC concentrations. With a uniform reduction in BC lifetime, the AeroCom Phase II models would strongly underestimate the near surface BC concentrations measured at Arctic stations, and moderately for airborne observations (Eckhardt et al. 2015).

Since wet deposition has been found to be one of the biggest sources of uncertainty, multiple studies have focussed on improving the model representation of the related microphysical processes or test the sensitivity of their respective models to them. For BC in the Arctic, improvements in the ageing, dry, and especially wet deposition parametrisation can improve the capability to reproduce Arctic BC concentrations in bulk microphysics models (e.g. GFDL-AM3, Liu et al. 2011). In this case a reduction of the wet scavenging by ice containing clouds by 95 % led to the biggest improvement in the vertical profile of Arctic BC concentrations. However, an improvement can also be achieved by introducing a microphysical approach instead of using an e-folding time for the ageing of BC.

Going from one to two-moments, tracking not only mass but also number concentration and separating the aerosol by modes, allows the consideration of the size dependence of the aerosol removal. This improves the ability of the model to reproduce measured profiles, compared to a bulk aerosol model (Vignati et al. 2010). This approach used by the M7 microphysics module by Vignati et al. (2004) is rare among the AeroCom Phase II models.

ECHAM-HAM uses M7 and has been shown to perform well in recent versions (Hodzic et al. 2019; Watson-Parris et al. 2019; Tegen et al. 2019). However, in terms of aerosol number concentrations, it shows a negative bias for the accumulation mode aerosol. In a sensitivity study perturbing single parameters, they show that this is potentially caused by an overestimation in wet deposition especially in the southern extratropics

Table 1.1.: BC lifetime in a range of current aerosol transport and climate models.

Model	Scheme	Ageing	Lifetime (d)	Wet dep. ($\mathrm{kt\,yr^{-1}}$)	Dry dep. ($\mathrm{kt\,yr^{-1}}$)	Study
TM5	modal	microphysics	6.2	8.0	0.2	Vignati et al. (2010)
GFDL-AM3	bulk	variable rates	9.5	4.4	3.3	Liu et al. (2011)
OsloCTM2	bulk	variable rates	6.3			Hodnebrog et al. (2014)
		tripled rate	3.9			
OsloCTM2	bulk	variable rates	6.0			Samset et al. (2014)
CAM4-Oslo	modal	coagulation only	8.2			Samset et al. (2014)
SPRINTARS	bulk		6.7			Samset et al. (2014)
ECHAM-HAM	modal	microphysics	5.5			Samset et al. (2014)
HadGEM2	modal	no	17.1			Samset et al. (2014)
GISS-modelE	bulk	fixed rate	5.9			Samset et al. (2014)
CAM5	modal	microphysics	3.8			Samset et al. (2014)
NCAR-CAM3.5	bulk	no	5.8			Samset et al. (2014)
GISS-MATRIX	bulk	microphysics	3.5			Samset et al. (2014)
GEOS-CHEM	bulk	fixed rate	4.2	5.0	1.5	Q. Wang et al. (2014)
GEOS-CHEM	bulk	microphysics	4.2	8.7	2.1	He et al. (2016)
GFDL-AM3	bulk	microphysics	7.7*	4.0	3.5	Shen et al. (2017)
CAM5-MAM4	modal	microphysics	4.3	/	/	Y. Wang et al. (2018)
		8 monolayers (from 3)	5.4	/	/	

(Watson-Parris et al. 2019).

1.4. book **outline**

As outlined before, the Arctic region is strongly affected by the current global climate change. Aerosol plays a potentially important role in this enhanced warming. BC, as a highly absorbing aerosol type in the visible light wavelength range, impacts the Arctic climate system when emitted locally or introduced by long-range transport. However, there are still large uncertainties in the climate impact of BC.

This work aims at quantifying the two key uncertainties of 1) the vertical distribution of BC and 2) the seasonality of BC. Modelling emissions and long-range transport have been identified as the main sources of uncertainty. It will provide a state-of-the-art estimate of the radiative impact of BC on the Arctic climate by facilitating global aerosol climate simulations.

The following scientific questions are answered:

- How does the Arctic BC concentration change under different emission representations?

- How does the wet removal of BC influence the long range transport of BC to the Arctic?

- What is the impact of BC on the Arctic's radiation balance?

To achieve these goals, the global aerosol-climate model ECHAM-HAM is used. The model performance is evaluated against a comprehensive set of surface measurements and vertical profiles. Two sensitivity studies are performed with regards to different BC emission representations and to a range of aerosol microphysical processes controlling the wet depostition of BC. For this BC tracers tagged by source region are introduced. The novel combination of these tagged tracer with the sensitivity studies on ageing and wet deposition gives new insights on transport pathways. Finally the radiative effects of BC in the Arctic climate system are assessed, for which the model was ungraded with the capability of excluding BC from the aerosol radiation interactions, allowing a better estimate of the direct radiative effect of BC than previously possible.

Chapter 2.

Methodology

2.1. The aerosol climate model ECHAM-HAM

2.1.1. ECHAM-HAM

ECHAM is the general circulation model developed at the Max Planck Institute for Meteorology (MPI-M) Hamburg (Stevens et al. 2013). Aerosol plays an important part in the dynamics of the atmosphere. To represent this influence, ECHAM is interactively coupled with the Hamburg aerosol module HAM (details in Zhang et al. 2012). The coupled ECHAM-HAM model, of which the version ECHAM6.3-HAM2.3 (Tegen et al. 2019) is used, is developed by the HAMMOZ community. For use in a global model, the number of aerosol types has to be restricted to keep the computational cost as low as possible, while still representing the most important processes. BC, sulphate (SU), OC, sea salt (SS), and mineral dust (DU) are the aerosol species considered in ECHAM-HAM.

Emissions

Volcanic emissions, biomass burning and anthropogenic emissions are prescribed from emission inventories. Biomass burning emissions are mixed into the boundary layer upon emission. The DU emissions from deserts as well as SS and dimethyl sulphide (DMS) from the ocean are calculated online, depending on the meteorology (see Zhang et al. 2012; Tegen et al. 2019). The DU emissions follow the scheme from Tegen et al. (2002) updated by a dust source activation frequency (Schepanski et al. 2007). SS is emitted following Gantt et al. (2011). DMS emissions from the ocean follow Nightingale et al. (2000).

M7: Aerosol micro physics module

The physical properties of aerosol are complex and depend on many factors, like their chemical composition, size, shape and temperature. These properties determine how the aerosol behaves in the atmosphere. The M7 micro physics module (Vignati et al. 2004; Zhang et al. 2012) is used in HAM to compute these properties. The aerosol particles in M7 are classified as either hydrophobic (often referred to as "soluble") or hydrophilic ("insoluble"), depending on their composition. M7 uses a "pseudomodal" approach to represent the size distribution of the aerosol particles. The size distribution in each mode is assumed to be log-normal. The prognostic variables in M7 are the aerosol number concentration for each of the seven modes, and the aerosol mass concentration for each mode and aerosol species in that mode.

The aerosol in the nucleation mode has a dry radius (r_{dry}) range of 0 nm–5 nm and a geometric standard deviation ($\sigma_{\ln r}$) of 1.59. It is always considered hydrophilic. The aerosol is separated between the hydrophilic and hydrophobic particles for the other modes: The Aitken mode (r_{dry}=5 nm–50 nm, $\sigma_{\ln(r)}$=1.59), accumulation mode (r_{dry}=50 nm–500 nm, $\sigma_{\ln(r)}$=1.59) and coarse mode (r_{dry} >500 nm, $\sigma_{\ln(r)}$=2.0).

M7: Particle ageing and nucleation

The BC, OC, and DU aerosol particles particles are emitted as hydrophobic and can later age to become hydrophilic. This is a prerequisite for nucleation scavenging, the most efficient wet removal pathway of these particles, as discussed in subsection 1.3.2. In M7 a particle is considered aged, when it is coated by at least one mono-layer of SU. The ageing happens via the pathways of condensation and coagulation.

The SU is produced by a simplified sulphur scheme considering a single production term, by the chemistry module. It is formed from SO_2 via a three-body reaction by oxidation with OH (Feichter et al. 1996). This available SU is then distributed by M7 via condensation on existing particles or nucleation of new particles.

All H_2SO_4 is removed from the gas phase in cloudy parts of the grid box by condensing it to the aerosol in the cloudy part, prioritizing the bigger aerosol modes, assuming that those are acting as cloud condensation nuclei.

In the rest of the grid box, the condensation sinks are calculated following the scheme by Fuchs (1964), with the condensation coefficient c_i of the mode i depending on the geometric mean radius $\bar{r}_i$, the diffusion coefficient D, mean thermal velocity v, the mean free path length of the SU molecule Δ', and an accommodation coefficient s. The nucleation coefficient is set to 0.3 for hydrophobic and 1.0 for hydrophilic aerosols:

$$c_i = \frac{4\pi D \bar{r}_i}{\frac{4D}{sv\bar{r}_i} + \frac{\bar{r}_i}{\bar{r}_i + \Delta'}} \tag{2.1}$$

The H_2SO_4 that remains is available for nucleation as SU particles into the nucleation mode. In the current model version, two different parametrisations are available. The first one by Vehkamäki et al. (2002) is derived from the classical nucleation theory. It can be used for temperatures between 230.15 K and 305.15 K and relative humidities between 0.01 % and 100 %. The second one by Kazil et al. (2007) is the default scheme. It utilizes a lookup table generated with a semi-analytical method, calculating the SU particle formation rate assuming a steady state and parametrised rate coefficients for the loss of gas molecules on aerosol particles.

The coagulation between particles is possible inside of the mode or between different modes. The scheme of Fuchs (1964) is used to calculate the coagulation coefficient K_{ij} between modes i and j, which takes averaged geometric mean radius $\overline{r}$ for both modes (Equation 2.3), the mean free path lengths Δ'' between the particles and thermodynamic state into account.

$$K_{ij} = \frac{16\pi \overline{D}\overline{r}}{\frac{4\overline{D}}{\overline{v}\overline{r}} + \frac{\overline{r}}{\overline{r}+\Delta''}} \tag{2.2}$$

$$\overline{r} = \frac{\overline{r}_i + \overline{r}_j}{2} \tag{2.3}$$

The thermodynamic state is represented through the averaged diffusion coefficient $\overline{D}$, and the averaged thermodynamic velocity $\overline{v}$. If the new radius is sufficiently large, the aerosol particles are moved to the corresponding mode. For the particles that coagulate with particles from a hydrophilic mode, the combined SU mass is calculated. From the new particle radius and the available SU mass the fraction of particles is calculated that can be covered with at least one monolayer of SU. This fraction is moved into the hydrophilic mode, the rest goes to the hydrophobic mode.

M7: Aerosol removal

The removal of aerosol from the atmosphere is split into three separate processes: the dry deposition, sedimentation and wet deposition. They are parametrised as functions of particle size, composition, mixing state and the surrounding meteorological conditions (Zhang et al. 2012). In ECHAM-HAM, the wet deposition is divided into a part that results from in-cloud scavenging of aerosol particles, and a second part that is scavenged by hydrometeors falling through the air below a precipitating cloud.

The below cloud scavenging is calculated separately for snow and rain (Croft et al. 2009). For both the fraction of aerosol particles colliding with a hydrometeor is calculated with the assumption that this results in collection of the particles. For rain, the scavenging ratio is dependent on the size distribution of the aerosol particles and rain droplets. For snow, there is only a dependence on the aerosol size distribution.

For the scavenging inside of stratiform clouds the diagnostic scheme described in Croft et al. (2010) is used. In this scheme, the collection of particles inside of the cloud is subdivided into the part that is acting as a CCN or an ice nucleating particle (INP) and another part which is colliding with droplets inside of the cloud. Only hydrophilic aerosol particles are assumed to act as CCNs and INPs.

Above 238.15 K, the number of aerosol particles removed (N) is equal to the cloud droplet number concentration (CDNC) and ice crystal number concentration (ICNC) calculated by the clouds physics of ECHAM-HAM. This available scavenging budget is divided among the hydrophilic aerosol modes (j) with radius greater than 35 nm. It is then partitioned according to the respective mode's aerosol number fraction among the total number of hydrophilic aerosol (with $r > 35$ nm):

$$N_{j,scav} = (\text{CDNC} + \text{ICNC}) \times \frac{N_{j,r>35\,\text{nm}}}{N_{r>35\,\text{nm}}} \tag{2.4}$$

With the total number of aerosol in the mode, the fraction of nucleated particles can be calculated.

$$r_{j,scav} = \frac{N_{j,scav}}{N_j} \tag{2.5}$$

The largest particles in each mode are assumed to form hydrometeors first. Therefore, the amount of aerosol mass that is removed corresponds to the biggest particles in the log-normal distribution. This is done by integrating over the tail of the distribution from the radius corresponding the smallest still nucleated particle. Then a mass fraction of nucleated aerosol mass can be calculated equivalently to Equation 2.5.

Below 238.15 K, homogeneous nucleation is assumed to be possible and therefore the nucleation scavenging is handled differently. A number of aerosol particles is scavenged anyway, but in this case all hydrophilic aerosol is scavenged starting from the largest mode (hydrophilic coarse mode), continuing down the modes by size. Once a mode is reached where the number concentration is higher than the remaining number of aerosol to be scavenged, the largest aerosol particles in this mode are scavenged, similar to the heterogeneous freezing nucleation scavenging.

Both hydrophilic and hydrophobic aerosol particles are scavenged by in-cloud impaction. The scheme used here is also described in Croft et al. (2010) and uses the same physical principles as the below-cloud impaction scavenging described by Croft et al. (2009). For cloud droplets it is size-dependent, utilising the distribution of the aerosol log-normal distribution, as well as gamma distribution of cloud droplets. For ice crystals the number concentration is assumed to be mono-disperse.

For convective clouds the removal of aerosol that act as CCN or INP is parametrised using prescribed scavenging ratios. These scavenging ratios change depending on aerosol mode and cloud type and were first introduced with an older scheme (details in Stier et al. 2005). The scavenging ratios are higher for hydrophilic than hydrophobic, and higher for larger particles.

The change in the mixing ratio C of tracer i by in-cloud scavenging is then:

$$\frac{\Delta C_i}{\Delta t} = P_i \cdot C_i \cdot f^{cloud} \cdot Q \tag{2.6}$$

where P_i is the sum of the scavenging ratios collected by nucleation and in-cloud impaction scavenging for both ice particles and liquid droplets, f^{cloud} is the cloud fraction of the grid box, and Q is the fraction of cloud water that is removed as precipitation.

Tagged tracers

To better understand the relation between Arctic BC concentration, its effect on the Arctic climate and the source regions, the BC tracers have been tagged with their source region. The chosen emission regions of Arctic, Russia, China, Europe, and North America can be seen in Figure 2.1. Additionally, the original tracer of BC remains for BC that has been emitted elsewhere in the world. This can be seen in Figure 2.2. It shows an averaged profile of BC mass mixing ratios for the Arctic (north of 65°N) in the year 2008. The coloured areas represent the BC from a run with the same setup as the base run, but with tagged tracers. The local Arctic emissions north of 65°N (light blue) are more prominent close to the surface than in the middle and upper troposphere. The higher altitudes are dominated by Chinese BC (red) and BC emitted in the untagged regions (black). The thick dashed green line shows the BC mass mixing ratio from the base run without tagged tracers, the thick orange line is the same setup as the base run but the same starting year as the tagged run (2007). It should be noted that, because numerical rounding effects, both of them are lower than the run with tagged tracers, showing that a comparison between a run with tagged tracers and one without should be taken with some caution.

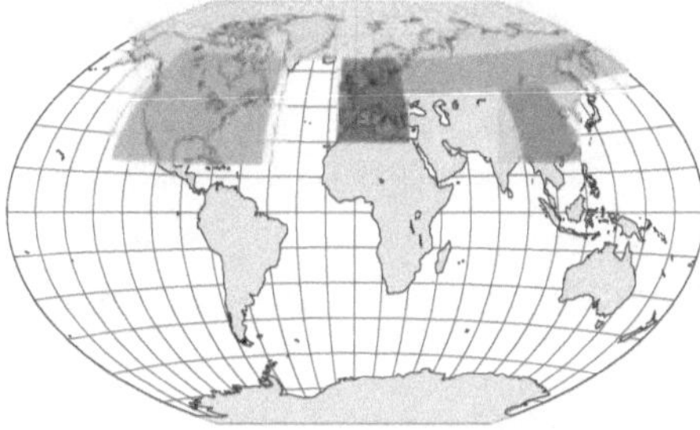

Figure 2.1.: Geographic regions where all emitted BC was tagged with the respective source region.

2.2. Comparison with observations

To evaluate the capability of ECHAM-HAM to reproduce the distribution of BC in the Arctic atmosphere and hence validate estimates of the DRE of BC, a set of aerosol concentration measurements is used. A combination of a long time series of measured

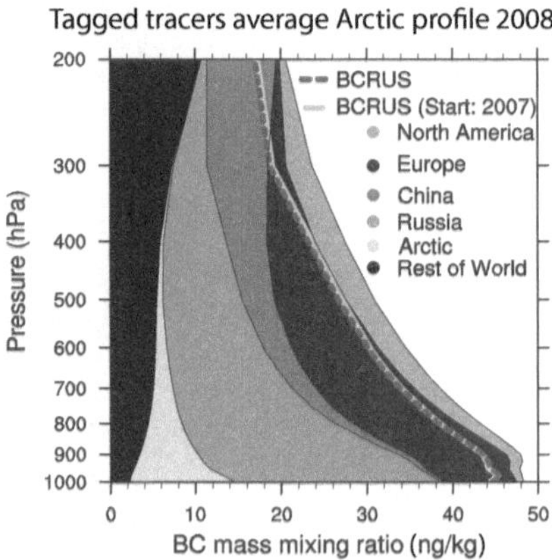

Figure 2.2.: Vertical profile of average Arctic (north of 65°N) BC for the year 2008 tagged by the source regions, as shown in Figure 2.1. The green dashed line shows the BC profile from the base run without tagged tracers. The orange line shows the profile of the same setup, but the same starting year as the tagged tracer run (later than *BCRUS*).

near surface concentrations at stations and airborne campaign measurements gives the best possible coverage of time and area.

In the Arctic, different measurement principles are used to provide an estimate of the BC concentration. These different measurement principles imply different definitions of BC.

The stations of the IMPROVE network use a filter analysis method with their own processing protocol (Chow et al. 2007). The measurements are of 24 h temporal resolution, which is lower than the other measurements, since each filter is analysed separately. The sample on a filter is volatilised, pyrolysed, then combusted and converted to CO_2, which then is measured with a laser. The measured quantity is EC, which is defined as a substance containing only carbon.

The Single Particle Soot Photometer (SP2) measures the laser-induced incandescence to derive refractory black carbon (rBC), which is defined as the carbonaceous fraction of particulate matter that is insoluble and vaporizes only at temperatures near 4000 K (Schwarz et al. 2010b).

The instruments and Aethalometer are both filter-based instruments that derive equivalent black carbon (eBC) from the light absorption of BC that has been collected on a filter. A mass absorption cross-section (MAC) can be used to derive the eBC concentration, which changes depending on the composition of the aerosol. Therefore, eBC is equal to the BC mass on the filter, if the composition and structure there is the same as when the MAC was measured. The definition of BC then has to be the same as the definition that is attached to the measurement method that was used to measure the MAC.

For this study a MAC of $9.8\,m^2\,g^{-1}$ is used for all datasets containing only the light absorption coefficient and no mass concentration or mass mixing ratio (Zanatta et al. 2018). This MAC is Arctic specific for aged aerosol at a wavelength of 550 nm. It was created with an SP2 as the reference method, therefore eBC equals to rBC where this MAC is used, with the additional uncertainty from the assumption of the mixing state. Zanatta et al. (2018) give an uncertainty range of $\pm 1.68\,m^2\,g^{-1}$, which translates to an uncertainty range of approximately $-20\,\%$ to $+15\,\%$ in eBC concentrations. The differences between measurement techniques and BC definitions are discussed in detail in Petzold et al. (2013)

The BC mass mixing ratios computed with ECHAM-HAM reference the carbon mass. Since this is the same idea as in the SP2, Particle Soot Absorption Photometer (PSAP) and Aethalometer measurements discussed, values from measurements and model will be compared without any further conversion. However, a comparison between a coarsely resolved model and point measurements is always challenging (Schutgens et al. 2016). They suggest the use of multiple independent measurements per grid box and temporal averaging. In this study, for each data point of measurements, a data point is taken from the 12-hourly model output that is collocated in space and time, as closely as the model resolution allows. For the station data we use the lowest model sigma layer and calculate the multi-year monthly median as well as the upper and lower quartiles of the BC mass mixing ratios.

To create vertical profiles from the model output, we define pressure bins into which the model data is sorted. Then, the median, upper and lower quartile are calculated from all data points in one pressure level bin for the BC mass mixing ratios. As the new vertical coordinate, the average pressure of all data points in that bin is used. When the different campaigns are compared in groups of seasons the median profiles of model and observation are averaged and the maximum and minimum among the medians are given.

2.2.1. Near-surface BC concentrations

Figure 2.3 shows the Arctic sites as triangles, where measurements of BC concentrations were taken. While the station data is only representative of the near surface air, the long time series gives robust information, in some cases spanning over a decade.

It is extremely valuable to understand the multi-year seasonality and the capability of ECHAM-HAM to reproduce this, since this has been discussed as one of the key

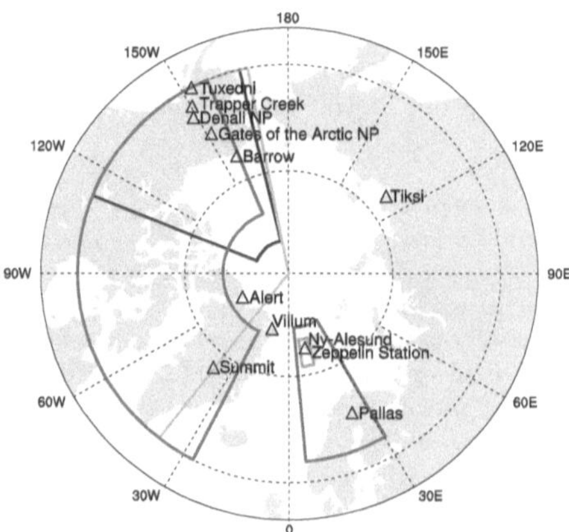

Figure 2.3.: Geographic regions of Arctic aircraft campaigns, the data which are used for model evaluation: Atmospheric Tomography Mission (ATom) in pink, HIAPER Pole-to-Pole observation (HIPPO) in blue, Arctic CLoud and Observations Using airborne measurements during polar Day (ACLOUD) and Polar Airborne Measurements and Arctic Regional Climate Model Simulation Project (PAMARCMiP)-2017 in green, Arctic Climate Change, Economy and Society (ACCESS) in red and Arctic Research of the Composition of the Troposphere from Aircraft and Satellites (ARCTAS) in orange. Black triangles show the location of stations with BC surface measurements. For the data sources see Table 2.2 and Table 2.1.

uncertainties of models (see section 1.3). The multi year monthly median and quartiles are chosen for the analysis of the seasonality over the average. It is the better tool than the average to gain information about the usual background state, because it is less influenced by peak concentrations during short events of high pollution, than the average.

To use the valuable information on the pollution events as well, additional analysis is performed. This data is used to analyse the temporal Pearson correlation with an emphasis on the spatial distribution of the respective sites. Because of the strong increases in BC concentration in case of a pollution event, this analysis is very sensitive to the capability of the model to time the event correctly. The capability to reproduce the exact concentration during one such event plays only a minor role. This is acceptable because any plume is "diluted" by spreading it out over the whole grid box, which is a known issue and not the focus of this analysis.

A detailed list of data providers, location, instrumentation and measurement period used in this study can be found in Table 2.1.

2.2.2. Vertical BC concentration profiles

Aircraft campaigns, while only representing temporally limited snapshots of the atmosphere, are the best way to evaluate the models' capabilities to produce correct vertical distributions, which is very important for producing reasonable values for the radiative effect of BC (Samset et al. 2013). Additionally, they cover big areas compared to stations, making them on average less dependent on local orographic features than station measurements. For the analysis of Arctic profiles only datapoints north of 60° N are used.

Table 2.2 gives an overview of the six campaigns used. All of them used SP2 instruments. Figure 2.3 shows the area covered by Arctic aircraft campaigns that measured BC concentrations as differently coloured boxes. The most western, eastern, southern (or 60°N) and northern extent are indicated by the edges.

The HIAPER Pole-to-Pole observation (HIPPO) campaign by the National Science Foundation (NSF) (S. C. Wofsy et al. 2017) is split into five deployments: HIPPO-1 (9 to 23 January 2009), HIPPO-2 (31 October to 22 November 2009), HIPPO-3 (24 March to 16 April 2010), HIPPO-4 (14 June to 11 July 2011) and HIPPO-5 (9 August to 8 September 2011). In the Arctic it covers wide areas of the American, Pacific Arctic and Bering Sea. As the name suggests, it additionally covers the Pacific region from pole to pole, down to 70°S. This is especially valuable for evaluating changes done to the aerosol microphysics that are affecting the concentrations not only in the Arctic. The aircraft used was the NSF/National Center for Atmospheric Research (NCAR) Gulfstream-V (GV).

The Arctic Research of the Composition of the Troposphere from Aircraft and Satellites (ARCTAS) campaign also had two deployments (data set – SP2_DC8; `https://www-air.larc.nasa.gov/cgi-bin/ArcView/arctas`, last access: 2 July 2018). ARCTAS spring in April 2008 and ARCTAS summer in June to July 2008. Flights were performed over North America and the American Arctic. It is important to note that this campaign was designed to measure biomass burning plumes in the Arctic, making it not representative of the usual Arctic BC profiles during spring and summer, but instead offering very interesting material for a case study. Jacob et al. (2010) describe the mission design in detail.

The ACCESS used the Falcon aircraft of the Deutsches Zentrum für Luft und Raumfahrt (DLR) to take BC concentration measurements over Scandinavia and the European Arctic (Roiger et al. 2015). Flights were performed in July 2012.

In March 2017, the BC concentrations were measured for an instalment of the Polar Airborne Measurements and Arctic Regional Climate Model Simulation Project (PA-MARCMiP) campaign. The profile used here was measured in March and was based in Longyearbyen, Spitzbergen, Norway. The SP2 was on board of the Polar 5 aircraft of the Alfred Wegener Institute (AWI).

The same aircraft was used during the ACLOUD campaign from 22 May till 28 June 2017 (Wendisch et al. 2018). It was based in Ny-Ålesund, Spitzbergen, Norway.

The ATom measured BC between July 2016 and May 2018 on the NASA DC-8 aircraft (S. Wofsy et al. 2018). All four installments are used here. Most Arctic measurements were taken over the American Arctic, as shown by the pink box in Figure 2.3. However, the campaign was performed between 90°N and 70°S, which is a great resource for the global evaluation. One of the most important features is that the flight tracks were designed with the goal to measure without a bias towards certain meteorological or aerosol conditions, making it especially well suited for a comparison with a global model. Many different variables were collected, of which here the BC, SU, SS and OC mass concentrations are used.

2.2.3. Aerosol Optical Thickness and Ångstrom Exponent

The aerosol optical thickness (AOT) as measured by the ground based sun photometers of the AErosol RObotic NETwork (AERONET, `http://aeronet.gsfc.nasa.gov`, last access: 10 June 2020, Holben et al. 1998), is used to evaluate the model performance globally and in terms of total aerosol. AERONET provides a globally distributed dataset, with excellent temporal coverage. When focusing on BC in the Arctic it is, however, only of limited use, because AERONET stations in the Arctic are relatively scarce, cannot measure during the Polar Winter, and BC is only one of the species that impact on the measured quantities. Nevertheless, the AOT can be used as a tool to compare the total amount of absorbing and scattering aerosol in the atmospheric column and the Ångstrom Exponent (AE) gives information on the average particle size of this aerosol. As in the previous model evaluation (Tegen et al. 2019), the model results are linearly interpolated to the time and location of the measurements, which they are compared against (Tegen et al. 2019). The 6 h averages of cloud-screened level 2 AOT data are used, as measured at 675 nm. The observational AE is derived from measurements of the extinction at 440 and 870 nm. The collocated modelled values of the AE are derived from AOTs at 550 and 865 nm.

2.3. Sensitivity Studies

As discussed in section 1.3, the emissions of BC, as well as the ageing and subsequent wet removal of BC, are the biggest sources of uncertainty for modelling the atmospheric transport to the Arctic and vertical distributions in the Arctic.

The amount and distribution largely govern the DRE of BC, which is of interest in the context of the AA. Therefore, two sets of sensitivity studies (one for emissions and one for the ageing and wet removal) were designed to span a range of reasonable assumptions. The results of the setups described here are later evaluated with observations

and analysed with an additional focus on the DRE of BC. Table 2.3 gives an overview
of the runs performed for these two sensitivity studies.

2.3.1. Emissions related uncertainties

As mentioned in section 2.1.1, ECHAM-HAM depends on emission inventories for BC,
OC, SU and its precursors. There are however many different emission inventories
available. They are often designed to fit a specific goal. A study that uses transient
simulations, where daily changing emissions are important, has different needs than a
model intercomparison project, where consistency among different models is desirable.
These different emission setups introduce an uncertainty into the calculation of the
effect of aerosol on the climate in general and therefore the Arctic climate as well.

To quantify the uncertainty on the Arctic burden and the related direct radiative
impact, a set of four different state-of-the-art emission inventories is used for this
manuscript (see also Schacht et al. 2019). Arctic averages are calculated for the region north of 60° N.

The emission datasets used include the emissions that were developed for the Atmospheric Chemistry and Climate Model Intercomparison Project (ACCMIP, Vuuren et
al. 2011). The horizontal resolution is $0.5° \times 0.5°$. It includes anthropogenic as well
as biomass burning emissions. There are historic emission data available until the year
2000, as well as projections for later years (2000 to 2100) that follow the Representative
Concentrations Pathways (RCPs) (Lamarque et al. 2010). The anthropogenic emissions
remain constant throughout the year. The biomass burning emissions vary monthly,
which is achieved by a scaling factor. Here, only the year 2000 emissions are used. The
ACCMIP emissions are still widely used for model experiments, in some cases using the
RCPs (Lund et al. 2018), in others using the fixed year 2000 emissions (Samset et al.
2014; Sand et al. 2017). Using fixed/harmonised emissions for one year is a common
simplification to reduce degrees of freedom and control the boundary conditions for
non-transient climate studies (e.g. Schulz et al. 2006; Schulz et al. 2009; Sand et al.
2017).

A different global emission data set of anthropogenic emissions is Evaluating the Climate and Air Quality Impacts of Short-lived Pollutants (ECLIPSE, Klimont et al.
2017). Here, version 5a (v5a) is used. The horizontal resolution is $0.5° \times 0.5°$. The
historic emissions cover the years until 2010. Projections for different industrial development scenarios linked to the RCPs are available for all further years. Here, the
scenario linked to the RCP4.5 is used for the years after 2010. Unlike the ACCMIP
emissions, the ECLIPSE emissions vary seasonally. This seasonality differs between
the sectors considered, by applying a sector specific annual cycle. It is important to
note that ECLIPSE includes gas flaring emissions. These have been discussed as being
important BC sources especially for the Arctic, but are considered to be difficult to
quantify (Stohl et al. 2013). In ACCMIP they are not considered at all, in other emis-

sion inventories they are possibly too low (Stohl et al. 2013; K. Huang et al. 2015).

K. Huang et al. (2015) created a highly resolved regional BC emission dataset for Russia, that not only takes gas flaring into account but assumes a higher emission factor than ECLIPSE. The emissions of Russian gas flaring are more than 40 % higher than those in ECLIPSE. The original horizontal resolution is $0.1° \times 0.1°$, but was scaled down to $0.5° \times 0.5°$. To cover the whole globe, it was combined with the ECLIPSE emissions. To generate the seasonal emission cycle the ECLIPSE sector specific factors were applied. This dataset represents a reasonable high estimate for Russian gas flaring emissions.

For most runs, daily changing biomass burning emissions are used (Kaiser et al. 2012). The dataset is named Global Fire Assimilation System (GFAS) and uses the fire radiative power as derived using the MODIS instrument aboard the NASA satellites Aqua and Terra. It derives not only the emission rate, but also the injection height into the atmosphere from the fire radiative power. This additional information is however not used by ECHAM-HAM, instead the aerosol is distributed through the boundary layer. Most of the time this is a reasonable approach for small and medium sized fires, but leads to an underestimation for very large and strong fires (Sofiev et al. 2009). Previous studies with ECHAM-HAM often multiplied the emissions of GFAS with a factor of 3.4 as suggested by Kaiser et al. (2012). This led to an overestimation in Arctic BC and is therefore not explored in this sensitivity study, but is shortly discussed in the context of the ageing and wet deposition related sensitivity study.

In order to study the effect these different inventories have, they were combined, and runs of at least 11 years were performed (for an overview of all runs see also Table 2.3):

- A run with only ACCMIP emissions, including anthropogenic as well as natural biomass burning emissions. They are fixed year 2000 emissions. The run represents a setup that can be used for non-transient model intercomparison studies. It is the only run not using the daily updated GFAS biomass burning emissions and is therefore not expected to reproduce biomass burning plumes. In comparison to other runs, this allows the estimation of the uncertainty related to the representation of biomass burning emissions. It is referred to as ACCMIP2000.

- In the run ACCMIP-GFAS the biomass burning emissions of ACCMIP were replaced with the daily changing biomass burning emissions of GFAS. It is otherwise unchanged from ACCMIP2000. The run therefore does not take changes in anthropogenic emissions into account. It can in conjunction with other runs thus be used to evaluate the uncertainty related to anthropogenic emissions.

- The global ECLIPSE v5a anthropogenic emissions were combined with the GFAS wildfire emissions in the run ECLIPSE.

- For the run BCRUS the anthropogenic emissions in Russia were replaced with the emissions by K. Huang et al. (2015). They were scaled with the monthly emission factors of ECLIPSE v5a to create a seasonal emission cycle. The rest

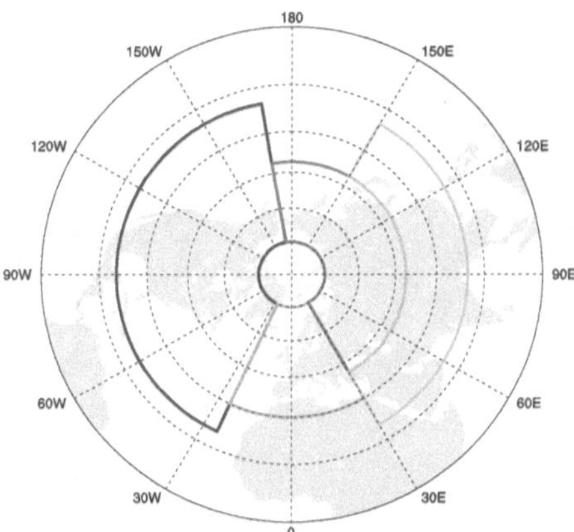

Figure 2.4.: Regions indicate the area used for averaging presented in Table 2.4. North America in blue, Europe in green, Russia in red and central Asia in orange.

remains the same as in the ECLIPSE run. This run has higher anthropogenic emissions in Russia than ECLIPSE. A lot of this increase is related to a higher emission factor for gas flaring, that has been assumed by K. Huang et al. (2015). In combination with other runs it gives an uncertainty range in Russian gas flaring emissions. Since Russian gas flaring has been discussed as possibly too low in current emissions inventories (Stohl et al. 2013), this setup is assumed to provide a good high estimate and is used as the reference run. In the second sensitivity study, concerning the effects of the representation of aerosol ageing and wet deposition, it is therefore named *Base*.

Figure 2.5a shows the emissions of BC for BCRUS. The highest emissions north of 30°N are found in the industrial regions of East Asia, Europe, northeastern America and in the gas and oil extraction centers in North America and northern Russia. The densely populated Europe produces much higher BC emissions than the sparsely populated Canada and Alaska. While the transport efficiency from East Asia can be assumed to be low, they are likely to have an important impact on the Arctic, because of the high emissions by long-range transport to the upper troposphere (Ikeda et al. 2017). The emissions for the four runs differ regionally, as can be seen in Table 2.4, with Figure 2.4 showing the areas referred to in the table, and locally as shown in Figure 2.5. As discussed the reference run (BCRUS) has the highest emission in Russia with $687\,\mathrm{kt\,yr^{-1}}$. The ECLIPSE run differs from BCRUS mostly in Russia, with $42\,\mathrm{kt\,yr^{-1}}$ less and small differences elsewhere, because the border of the country differs the red box in Figure 2.4. The change in emissions between these two runs can be seen in Figure 2.5c. Both of them show much higher emissions in Central Asia than any of

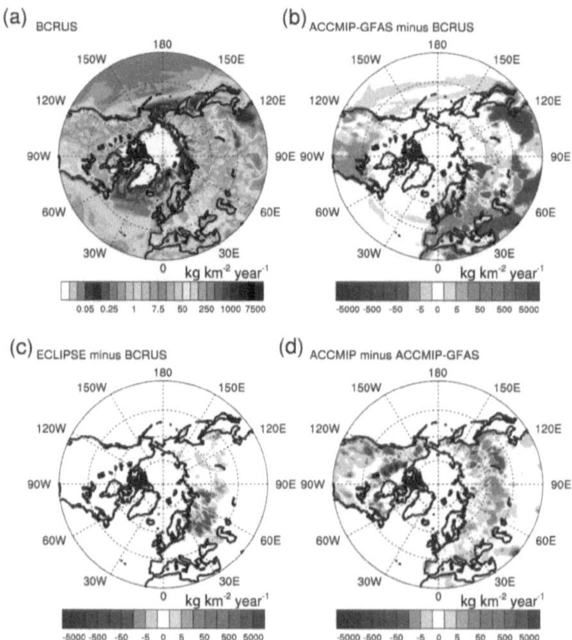

Figure 2.5.: Maps of annual mean BC emissions for the years 2005–2015. (a) Absolute values are given for BCRUS. Difference between (b) the ACCMIP-GFAS and BCRUS results, (c) the ECLIPSE and BCRUS results and (d) between the ACCMIP and ACCMIP-GFAS results.

the runs utilizing the ACCMIP year 2000 emissions (ACCMIP2000/ACCMIP-GFAS) by almost $1000\,\mathrm{kt\,yr^{-1}}$. The emissions in North America and Europe are lower by over $100\,\mathrm{kt\,yr^{-1}}$ for both regions. This pattern can be seen in Figure 2.5b which shows the difference in BC emissions between ACCMIP-GFAS and BCRUS. It is caused by the development in the regions caused by the economic growth in central Asia and by recent changes in air quality regulations in Europe and North America. The change in the biomass burning representation from ACCMIP2000 to ACCMIP-GFAS leads to higher emissions in North America by $65\,\mathrm{kt\,yr^{-1}}$ and lower emissions by $36\,\mathrm{kt\,yr^{-1}}$ in Russia. Locally the changes are less uniform as can be seen in Figure 2.5d, caused by the regions of biomass burning changing from year to year in ACCMIP-GFAS while they repeat each year in ACCMIP2000.

2.3.2. Sensitivity to ageing and removal

The wet removal of aerosol is the primary aerosol particle sink in ECHAM-HAM (Croft et al. 2010). For BC the in-cloud nucleation scavenging is more efficient than impaction scavenging. For particles that are hydrophobic upon emissions like BC, a particle ageing is a prerequisite for nucleation scavenging (Croft et al. 2010), as described in section 2.1.1. Since Arctic BC is in most cases transported from far away sources over long time scales, it is both subject to longer ageing, as well as more likely to interact with rain.

The wet deposition flux adjustment is done by multiplying the tracer changes from in-cloud scavenging in each time step by a factor α_{ic}. Equation 2.6 thus changes to:

$$\frac{\Delta C_i}{\Delta t} = P_i \cdot C_i \cdot f^{cloud} \cdot Q \cdot \alpha_{ic} \tag{2.7}$$

There is a large uncertainty in the number of SU monolayers surrounding an otherwise hydrophobic particle that are needed to sufficiently age it to a point where it can be considered hydrophilic. Aerosol climate models use a range of different values as their default (the CAMS model uses three monolayers) and it is considered a tunable parameter (Y. Wang et al. 2018). This number of monolayers needed is strongly connected with the ageing speed. A number of sensitivity studies have therefore been performed that take this factor among others into account (e.g. Lohmann et al. 2009; Y. Wang et al. 2018; Watson-Parris et al. 2019). Here, this threshold number, which is set to unity in the base version of ECHAM-HAM, is multiplied by a scaling factor α_{ml}.

Since the Arctic climate system is unique and most Arctic aerosol is transported there by long range transport, a set of sensitivity studies has been performed to study the effects that control the removal of the aerosol:

- The wet deposition flux has been increased by multiplying the in cloud scavenging by a factor of two. This simulates a higher scavenging of the aerosol particles in the cloud by both nucleation and impaction scavenging. It is referred to as *2xINCLOUD*.

- A decrease in the scavenging by a factor of two, to simulate a lower scavenging, referred to as *0.5xINCLOUD*.

- An increase in the number of SU monolayers that are required to age a particle from the default of one to five monolayers. This means it takes more available SU to age an particle, which slows down the process. Since the aerosol from the hydrophilic modes is removed more efficiently by wet deposition this affects the lifetime. The experiment is referred to as *5xMLAYERS*.

- The experiment with a lower ageing speed, meaning 0.3 SU monolayers, is referred to as *0.3xMLAYERS*.

- An increase in SU abundance is tested in *2xSU*, where the global emissions of SU and its precursors are doubled.

- In *0.5xSU* the global emissions of SU are instead decreased by a factor of two.

- A combination of a higher in-cloud scavenging ($\alpha_{ic} = 2$) but slower ageing ($\alpha_{ml} = 5$) is tested in *Combined$_{ic=2,ml=5}$*.

The corresponding runs are limited to two years to save computational resources. These changes are still within the range of what can be considered realistic (Y. Wang et al. 2018; Watson-Parris et al. 2019). It has to be noted that these changes will directly affect all aerosol types and not only BC.

As discussed later, ECHAM-HAM does show a high sensitivity to changes in α_{ic}, and α_{ml}. A slower ageing, meaning a higher required number of SU monolayers for ageing, combined with a more effective in-cloud scavenging are the most promising way to tackle issues that the evaluation reveals. With the results from this first set in mind, a second set of sensitivity studies is performed to find a setup which improves the capabilities of ECHAM-HAM to produce observed Arctic vertical BC profiles as well as global observations. For this α_{ml} and α_{ic} are both changed as in the *Combined* run but with varying levels of strength. The resulting setups are named in the convention of *Combined$_{ic=i,ml=j}$*, where $i = \alpha_{ic}$ is the factor the in-cloud scavenging is enhanced by, and $j = \alpha_{ml}$ is the number of required monolayers to age a particle. The runs are:

- *Combined$_{ic=3,ml=5}$*: $\alpha_{ic} = 3$, $\alpha_{ml} = 5$

- *Combined$_{ic=2,ml=3}$*: $\alpha_{ic} = 2$, $\alpha_{ml} = 3$

- *Combined$_{ic=2,ml=2}$*: $\alpha_{ic} = 2$, $\alpha_{ml} = 2$

- *Combined$_{ic=1.5,ml=2}$*: $\alpha_{ic} = 1.5$, $\alpha_{ml} = 2$

2.4. Radiative effects of BC

The radiative balance of the atmosphere drives the climate system. The impact of BC (or any other aerosol) on the climate is investigated in terms of atmospheric radiative effects and the BC-in-snow albedo effect, with a focus on the Arctic climate. For Arctic averages the area north of 60° N is considered. The radiative effect of BC is calculated as the difference in net (downward minus upward) irradiance caused by BC. Here, it is defined as the total of the solar and thermal wavelength range. The effective radiative effect (ERE) is defined following the Intergovernmental Panel on Climate Change (IPCC) as the change in net downward radiative flux density (F^{net}) at the TOA, allowing a readjustment of the atmospheric temperatures, water vapour, and clouds, but keeping the sea-ice concentrations and sea surface temperatures fixed (Myhre et al. 2013a). The ERE (ΔF^{net}) is thus given for the total aerosol as:

$$ERE_{all\ aerosol} = \Delta F^{net}_{aerosol} = F^{net}_{base} - F^{net}_{no\ aerosol} \tag{2.8}$$

And for BC (ΔF^{net}_{BC}) as:

$$ERE_{BC} = \Delta F^{net}_{BC} = F^{net}_{base} - F^{net}_{no\ BC} \tag{2.9}$$

The ERE is composed of the direct radiative effect (DRE), the semi-direct radiative effect (sDRE), and indirect radiative effect (IRE):

$$ERE = DRE + sDRE + IRE \tag{2.10}$$

The DRE is in some cases referred to as "instantaneous radiative forcing". In the literature sometimes a distinction is made between "forcing" and "radiative effect", where the "forcing" only describes the impact by the change in concentration from pre-industrial levels, and "radiative effect" describes the impact of the total concentration. The latter is used here.

The IRE describes the change in the radiation budget caused by changes in cloud properties (lifetime, cloud cover) caused by direct interaction between cloud and aerosol particles on a microphysical level (i.e. the ability of aerosol particles to act as CCN and INPs). The sDRE describes the effect of cloud property differences caused by changes in the thermodynamic environmental conditions for the cloud and the effect of a stabilisation of the atmospheric stratification. For BC the sDRE generally means an absorption of solar radiation, increasing the air temperature which causes a burn-off of clouds and a stabilisation of the atmospheric stratification.

As described in section 1.2 in more detail, these effects change with time of year and environmental conditions. As in the other chapters, the focus will be on BC and specifically for the Arctic, which means north of 60° N unless stated othewise.

2.4.1. Atmospheric radiative effects

Since BC is known to be highly absorbing in the visible light spectrum, the DRE of BC is the main focus in terms of radiative effects for this work. Up to this point, there has been no option included in ECHAM-HAM to calculate the DRE of a single aerosol type. However, ECHAM-HAM has the capability to calculate the TOA and bottom of the atmosphere (BOA) instantaneous DRE of all aerosol. This is done by calling the radiative scheme twice, once with and once without considering aerosol particles, optically. This second call has no influence on the dynamics of the model, meaning that e.g. clouds are not changed. From the difference between the radiative balances of both calls the DRE of all aerosol types included ($DRE_{all\ aerosol}$) can be calculated:

$$DRE_{all\ aerosol} = \Delta F^{net}_{all\ aerosol,\ direct} = F^{net} - F^{net}_{transparent\ aerosol} \qquad (2.11)$$

The combined DRE of all considered species is therefore available as model output from the single run "BCRUS (Base)".

To properly calculate the DRE of only BC, the model code was upgraded, to allow a separate run "BCRUS-TRBC", where BC does not interact with radiation (is "transparent"), but is still available for aerosol cloud interactions as well as interaction with other aerosols. This is achieved by skipping BC in the calculation of the complex refractive index and adjusting the number concentration of particles that is used for the calculation of the total aerosol optical thickness. However, the wet radius of the aerosol modes including BC, that goes into the calculation of the refractive index, was not adjusted. This additional run (BCRUS-TRBC) yields:

$$DRE_{non-BC\ aerosol} = \Delta F^{net}_{non-BC\ aerosol,\ direct} = F^{net}_{all\ but\ BC} - F^{net}_{transparent\ aerosol} \qquad (2.12)$$

where $F^{net}_{all\ but\ BC}$ is the net radiative flux density of everything, excluding the direct interaction of BC with radiation and $F^{net}_{transparent\ aerosol}$ is the net radiative flux density from the second call of the radiation scheme of the same run (BCRUS-TRBC) that is skipping aerosol.

The DRE of BC then simply is the difference from DREs of the runs BCRUS (Base) and BCRUS-TRBC:

$$DRE_{BC} = DRE_{all\ aerosol} - DRE_{non-BC\ aerosol} \qquad (2.13)$$

Excluding BC from the direct interaction with radiation is the only feasible way of calculating the DRE of BC without strongly impacting other aerosol populations, causing a considerable error in the estimate of the DRE of BC. It should be noted that with this method, the estimate does include a small second order effect on the DRE that is caused by the semi-direct effect of BC between the two different runs. Since the sDRE of BC is already small on the large-scale average, because of positive and negative effects on the cloud formation that cancel each other out (Tegen et al. 2018), this second order effect will be considered negligible.

The semi-direct radiative effect of BC is the change in F^{net} through changes in clouds, caused by the atmospheric dynamic feedback due to the DRE of BC. It can therefore be calculated from the runs used for calculating the DRE of BC, BCRUS and BCRUS-TRBC:

$$sDRE_{BC} + DRE_{BC} = F^{net}_{normal} - F^{net}_{transparent\ BC} \qquad (2.14)$$

Where F^{net}_{normal} and $F^{net}_{transparent\ BC}$ are the net radiative flux density of BCRUS (Base) and BCRUS-TRBC, respectively.

The IRE of BC from both runs cancel each other out. They therefore do not show up in Equation 2.14. From that follows:

$$sDRE_{BC} = F^{net}_{normal} - F^{net}_{transparent\ BC} - DRE_{BC} \tag{2.15}$$

The only way of properly estimating the IRE of BC is to prohibit aerosol-cloud interactions in the model (Cherian et al. 2017). Setting up the model for this method is outside of the scope of this book. There is, however, another method that allows a rough estimate. To calculate the IRE with this method, another additional run (BCRUS-NOBC) is required. In this run no BC is emitted, making sure no interaction between cloud and BC is possible. The difference in F^{net} between the run with BC and the run without BC then gives F^{net}_{BC} (the ERE of BC):

$$DRE_{BC} + IRE_{BC} + sDRE_{BC} = F_{BC} = F^{net}_{normal} - F^{net}_{no\ BC} \tag{2.16}$$

By reformulating Equation 2.16 it is possible to calculate the IRE of BC:

$$IRE_{BC} = F^{net}_{normal} - F^{net}_{no_BC} - DRE_{BC} - sDRE_{BC} \tag{2.17}$$

However, the estimate suggested by Equation 2.17 contains also the effect that omitting BC from the model has on the concentration of other aerosols. This change in other concentrations again influences the radiative budget (through direct aerosol radiation interaction and cloud-mediated). The estimate in Equation 2.16 is, therefore, already erroneous in that it omits second order effects.

One of the second order effects is a change in DRE of the total aerosol caused by a difference in the concentration of non-BC aerosol between BCRUS (Base) and BCRUS-NOBC. It is a pseudo direct radiative effect (pDRE) of BC, caused by BC, but neither through direct interaction of BC with radiation, nor through interaction of BC with clouds. This effect does not appear in Equation 2.13, but does if BCRUS-NOBC is used equivalently:

$$DRE_{BC} + pDRE_{BC} = DRE_{all\ aerosol} - DRE_{no\ BC} \tag{2.18}$$

It follows that the $pDRE_{BC}$ can be calculated from the runs BCRUS-TRBC and BCRUS-NOBC:

$$pDRE_{BC} = DRE_{BC,\ transparent\ BC} - DRE_{BC,\ no\ BC} \tag{2.19}$$

Additionally there are also changes to the IRE_{BC} and $sDRE_{BC}$, caused by the omission of BC. Estimating the pseudo semi-direct radiative (psDRE) and pseudo indirect radiative effect (pIRE) would require following the methodology of Cherian et al. (2017). Doing this is, however, out of the scope of this work.

The pDRE of BC can be used as a correction, getting closer to the estimate of Cherian et al. (2017). From Equation 2.17 and Equation 2.19 follows:

$$IRE_{BC} = F^{net}_{normal} - F^{net}_{no_BC} - DRE_{BC} - sDRE_{BC} - pDRE_{BC} \tag{2.20}$$

2.4.2. Radiative effects of BC in snow

As described in section 1.2 BC does not only have an effect on the radiation while suspended in the air, but also when deposited on snow and ice. ECHAM-HAM is capable of calculating this effect, either as a diagnostic only, or allowing interaction with the model dynamics. The BC-in-snow albedo effect is parameterized using a lookup table, that has been generated with a single-layer version of the Snow, ice and Aerosol Radiation (SNICAR) model (details in M. G. Flanner et al. 2007). From the snow precipitation, the deposition flux of BC, snow melt and glacier runoff, the BC concentration in the top 2 cm is calculated. The snow melt and glacier runoff can cause an enrichment of BC.

In the current version, the calculated changed surface albedo only affects solar radiation. The effect of BC on the terrestrial spectrum is, however, small. Additionally the effect of BC deposited on bare sea ice is not considered. This effect is however negligible, because the spacial extent of bare sea ice is small (Gilgen et al. 2018). In the model runs used here, the BC-in-snow albedo effect does not influence the model dynamics.

2.5. Statistical Significance

To be able to evaluate the significance of the sensitivity studies a special criterion suitable for spatially correlated data is used (Wilks 2016). This is needed, since the usual procedure is to apply a two sided Student's t-test on whether the two fields from two different sensitivity runs of the same variable are significantly different. For this a probability p is chosen at which the null hypothesis H_0 (that both fields are from the same population) is rejected, even though they actually are significantly different. Since this test is done for all grid boxes separately, this means that with the commonly chosen value of $p = 0.05$ every twentieth grid box would on average be falsely labelled as significantly different between the two experiments.

The criterion controls this false discovery rate (FDR) by defining smaller local values p_i, where $i = 1, ..., n$ for n local hypothesis tests. The p_i are sorted in ascending order. A threshold value $p^\star_{\mathrm{FDR}}$ is then defined, below which H_0 is rejected for each local test:

$$p^\star_{\mathrm{FDR}} = \max_{i=1,...,n} \left[p_{(i)} : p_{(i)} \leq (i/n)\, \alpha_{\mathrm{FDR}} \right] \tag{2.21}$$

Here α_{FDR} is the chosen control level for the FDR. It is set to $\alpha_{\mathrm{FDR}} = 0.05$ in this book.

In most cases a significant difference according to the aforementioned criterion is calculated when two fields of the same variable, but from different model runs are compared. However, in some cases it is used to show whether a radiative effect significantly alters a different radiative effect. It is, for example, tested whether the DRE of BC significantly alters the DRE of all aerosol, by comparing the DRE field of a run with and a run without BC-radiation-interaction. For the DRE it is tested whether the TOA net radiation balance of the monthly means is significantly different between the run with BC considered by the radiation scheme and in the run with BC that does not interact with the radiation scheme.

Table 2.1.: Station measurements overview. Period describes the period used in this work.

	Latitude	Longitude	Period	Instrument/Inlet	Reference
Alert	82.492°N	62.508°W	01-2012 - 12-2014	Aethalometer/total	Backman et al. (2017)
Pallas	67.973°N	24.116°E		Aethalometer/total	
Tikang	71.973°N	128.889°E		Aethalometer/PM10	
Summit	72.580°N	38.480°E		Aethalometer/PM2.5	
Zeppelin	78.907°N	11.889°E		Aethalometer/total	
Ny-Ålesund	78.927°N	11.927°E	04-2012 - 12-2015	PSAP/PM10	Sinha et al. (2017)
Barrow	71.288°N	156.792°W	08-2012 - 12-2015	PSAP/PM10	
Villum	81.600°N	16.667°W	05-2011 - 08-2013	MAAP/total	Massling et al. (2015)
Gates of the Arctic	66.903°N	151.517°W	11-2008 - 10-2015	Quartz-fibre filter/PM2.5	Chow et al. (2007)
Denali	63.723°N	148.968°W	01-2005 - 12-2015	Quartz-fibre filter/PM2.5	Chow et al. (2007)
Trapper Creek	62.315°N	150.315°W	01-2005 - 12-2015	Quartz-fibre filter/PM2.5	Chow et al. (2007)
Tuxedni	59.992°N	152.666°W	01-2005 - 12-2014	Quartz-fibre filter/PM2.5	Chow et al. (2007)

Table 2.2.: Overview of the aircraft measurements. Period describes the period used in this work. For aircraft campaigns the location of the airfield is given, unless no specific base can be defined (denoted by *).

Campaign	Lat/Lon	Period	Reference
ACCESS	69.3°N/16.1°E	06-2012	Roiger et al. (2015)
ARCTAS	64.8°N/147.9°W	04-2008/07-2008	Yutaka Kondo
HIPPO 1-5	*	01-2009/09-2011	S. C. Wofsy et al. (2017)
ATom 1-4	*	08-2016/05-2018	S. Wofsy et al. (2018)
PAMARCMiP	78.2°N/15.5°E	03-2017	Herber et al. (2012)
ACLOUD	78.2°N/15.5°E	05-2017/06-2017	Wendisch et al. (2018)

Table 2.3.: Model run overview. The start of the period is always after a three months spin-up period. * Denotes that the code has changed to accompany the sensitivity study, but parameters remain the same.

Run	Period	Emissions		Microphysics	Tagged BC	Adv. Rad. Diag.
		anthropogenic	wild fires			
BCRUS (Base)	2005–2018	BCRUS	GFAS	unchanged	no	no
BCRUS-TRBC	2005–2018	BCRUS	GFAS		no	yes
BCRUS-BCTAG	2007–2008	BCRUS	GFAS		yes	no
ECLIPSE	2005–2015	ECLIPSE	GFAS		no	no
ECLIPSE-TRBC	2005–2015	ECLIPSE	GFAS		no	yes
ACCMIP-GFAS	2005–2015	ACCMIP	GFAS		no	no
ACCMIP-GFAS-TRBC	2005–2015	ACCMIP	GFAS		no	yes
ACCMIP2000	2005–2015	ACCMIP	ACCMIP		no	no
ACCMIP2000-TRBC	2005–2015	ACCMIP	ACCMIP		no	yes
2xINCLOUD	2007–2008	BCRUS	GFAS	$\alpha_{ic} = 2$	yes	no
0.5xINCLOUD	2007–2008	BCRUS	GFAS	$\alpha_{ic} = 0.5$	yes	no
5xMLAYERS	2007–2008	BCRUS	GFAS	$\alpha_{ml} = 5$	yes	no
0.3xMLAYERS	2007–2008	BCRUS	GFAS	$\alpha_{ml} = 0.3$	yes	no
2xSU	2007–2008	BCRUS (SU×2)	GFAS (SU×2)	unchanged*	yes	no
0.5xSU	2007–2008	BCRUS (SU×0.5)	GFAS (SU×0.5)	unchanged*	yes	no
Combined$_{ic=2,ml=5}$	2007–2011	BCRUS	GFAS	$\alpha_{ic} = 2$, $\alpha_{ml} = 5$	yes	no
Combined$_{ic=3,ml=5}$	2007–2008	BCRUS	GFAS	$\alpha_{ic} = 3$, $\alpha_{ml} = 5$	no	no
Combined$_{ic=2,ml=3}$	2007–2008	BCRUS	GFAS	$\alpha_{ic} = 2$, $\alpha_{ml} = 3$	no	no
Combined$_{ic=2,ml=2}$	2007–2008	BCRUS	GFAS	$\alpha_{ic} = 2$, $\alpha_{ml} = 2$	no	no
Combined$_{ic=1.5,ml=2}$	2007–2008	BCRUS	GFAS	$\alpha_{ic} = 1.5$, $\alpha_{ml} = 2$	no	no
BC-Optimised	2007–2018	BCRUS	GFAS	$\alpha_{ic} = 2$, $\alpha_{ml} = 2$	no	no
BC-Optimised-BCTAG	2007–2008	BCRUS	GFAS	$\alpha_{ic} = 2$, $\alpha_{ml} = 2$	yes	no
BC-Optimised 3.4xBC	2007–2018	BCRUS	GFAS×3.4	$\alpha_{ic} = 2$, $\alpha_{ml} = 2$	no	no

Table 2.4.: Area-weighted totals of BC emissions from anthropogenic sources and biomass burning fires for the main source regions (as shown in Figure 2.4) averaged for the years 2005–2015 in $2\,\mathrm{kt\,yr}^{-1}$.

Model Run	North America	Europe	Russia	Central Asia
BCRUS	400	408	687	2981
ECLIPSE	399	401	645	2983
ACCMIP	450	538	578	2005
ACCMIP-GFAS	515	533	542	1997

Table 2.5.: Different parts of the ERE of BC and how they can be obtained from the different model simulations. In some cases including 2nd order effects introduced by changes in the total aerosol composition, by omitting BC in BCRUS-NOBC (denoted by *).

Simulations	Variable	Radiative effect
BCRUS	F (double call)	DRE (total aerosol)
BCRUS		BC-in-snow albedo effect
BCRUS−BCRUS-TRBC	DRE	DRE of BC
BCRUS−BCRUS-TRBC	F	sDRE+DRE (both of BC)
BCRUS−BCRUS-NOBC	F	IRE+sDRE+DRE (all of BC)*

Chapter 3.

Model evaluation

The use of the measurements described in section 2.2 allows the evaluation of the capabilities of ECHAM-HAM to reproduce measured BC concentrations. With the long-term station measurements the quality of the modelled seasonal cycle can be assessed, and the airborne in-situ measurements are the best source of information about the vertical distribution, both of which are crucial for correctly estimating the radiative impact. Here, the emission setup *BCRUS*, as described in subsection 2.3.1, is used as the base run.

3.1. Near-surface BC mass concentration

The seasonality of BC concentrations is one of the key uncertainties in simulating Arctic BC (Samset et al. 2014; Arnold et al. 2016). Here, ECHAM-HAM is evaluated by comparing against measurements of stations located across the Arctic. The datasets span many years, some over a decade, allowing an assessment representative of their respective locations. A detailed list of the stations used can be found in Table 2.1. See also Figure 2.3 for their geolocation.

For each measured data point, a collocated data point is sampled from the model. Then the multi-year monthly median and quartiles of the near-surface BC concentrations were calculated for each of these stations. Figure 3.1 shows the resulting seasonal cycles. In general, the model is able to reproduce the monthly values with a maximum in the beginning of the year and the minimum concentrations in summer.

Zeppelin Station and Ny-Ålesund are both located on Svalbard (Figure 2.3). They are shown in Figures 3.1a and b, respectively. For Zeppelin Station, the peak median BC concentrations are found with $36\,\mathrm{ng\,m^{-3}}$ in March and slightly later for Ny-Ålesund with $30\,\mathrm{ng\,m^{-3}}$ in April. ECHAM-HAM overestimates the BC concentrations in the beginning of the year for both stations. In Ny-Ålesund the peak of the modelled median BC concentrations is found in February at $120\,\mathrm{ng\,m^{-3}}$ compared with an observed maximum of $20\,\mathrm{ng\,m^{-3}}$. For both stations the BC concentrations are also overestimated in November and December. The model simulates monthly medians in December of $90\,\mathrm{ng\,m^{-3}}$ for both stations, while observations only show $10\,\mathrm{ng\,m^{-3}}$ and $20\,\mathrm{ng\,m^{-3}}$ for

Zeppelin Stations and Ny-Ålesund, respectively. Since the modelled seasonal cycles at these two and other Arctic stations discussed later, are more similar to each than the corresponding observed ones, it is clear that the conditions in the Arctic are too uniform in the relatively coarsely resolved model. Wet scavenging and transport patterns have a large impact on the seasonal cycles (**sha13**; Browse et al. 2012; Sinha et al. 2017). The differences in precipitation in the winter months between the stations on Svalbard and those discussed later is, however, well reproduced. The overestimation is therefore more likely related to the latter. It has to be noted, that at the model resolution of approximately 1.8°, Zeppelin Station and Ny-Ålesund are located in the same grid box. The difference in altitude is not taken into account from the model side; instead the lowest level above the modelled orography is chosen. Differences in the model values of the BC concentration between the two stations, shown in Figure 3.1, are therefore only due to the different temporal availability of the measurements. The agreement between model and measurement is slightly better for Zeppelin Station. Its location makes it more exposed to long-range transport, while Ny-Ålesund is often subject to a blocking situation that prevents mixing of air masses, because it is located inside a fjord.

Figures 3.1c and d show the Villum Research Station and Alert that are both located in the north of Greenland. For them the agreement between model and measurement is better than for the Svalbard stations. They show a similar pattern of a maximum in observed concentrations in February with $70\,\mathrm{ng\,m^{-3}}$. The modelled peak BC concentration occur earlier in the year and higher with $70\,\mathrm{ng\,m^{-3}}$ and $100\,\mathrm{ng\,m^{-3}}$ for Alert and Villum Research Station, respectively.

The seasonal cycle for Summit, which is located on top of the Greenland ice sheet, is different, see Figure 3.1e. The median BC mass concentrations in Summit are very low. The highest median BC mass concentrations were observed in April with slightly more than $30\,\mathrm{ng\,m^{-3}}$. The annual cycle is not very pronounced for the station. In summer, concentrations are slightly lower, but the minimum was observed in January, when ECHAM-HAM, however, is computing the highest median BC concentrations. Even though the model is incapable of producing this seasonal cycle, the amount of BC agrees well between model and observations, with values generally below $30\,\mathrm{ng\,m^{-3}}$.

For Barrow the annual cycle, shown in Figure 3.1f, is reproduced well. However, ECHAM-HAM overestimates the BC concentration between October and March. The highest median is correctly computed for February, but is too high with $100\,\mathrm{ng\,m^{-3}}$ compared with the observation of $60\,\mathrm{ng\,m^{-3}}$.

The annual cycle for Tiksi is very well reproduced by ECHAM-HAM, as can be seen in Figure 3.1g. It is the only available station in the Russian Arctic and shows the highest observed median concentrations among these stations. The highest median concentration is observed in February with $200\,\mathrm{ng\,m^{-3}}$. This is only slightly underestimated by the model with $140\,\mathrm{ng\,m^{-3}}$. Even though only slightly, ECHAM-HAM underestimates near-surface BC in all months that are showing median BC concentrations of more than $50\,\mathrm{ng\,m^{-3}}$ in the observations.

The pattern for Pallas, shown in Figure 3.1h, is similar to Barrow and Tiksi, with high concentrations in winter and early spring and low concentrations in summer. However, concentrations are higher than in Alert and Villum, with a modelled peak median concentration of $300\,\mathrm{ng\,m^{-3}}$ in January, which is the highest among the Arctic stations, and $140\,\mathrm{ng\,m^{-3}}$ in March.

The last four stations compared here are all located in Alaska (see Figure 2.3) and belong to the IMPROVE network. They are shown in Figures 3.1i through l. For all of them, the highest concentrations are observed in summer. The lowest are found in winter. ECHAM-HAM is not reproducing this behaviour and instead shows a seasonal cycle that is comparable to Villum and Alert.

However, the modelled maximum appears one month too early, in March. The modelled near-surface concentrations are too high for autumn and winter in the Central Arctic. The separate annual cycles observed and modelled for these stations can be found in Figure 3.1). In general, the seasonality in near-surface BC concentrations is reproduced well by ECHAM-HAM. Possibly missing emission alone would only explain the underestimation in summer, but not the overestimation in winter. The Brooks Range spans through Alaska, from the Bering Sea in the west to the Beaufort Sea in the east, with multiple peaks of more than $2000\,\mathrm{m}$ above sea level. The IMPROVE stations are all located south of these mountains. An underestimation of the orographic height in the coarsely resolved model leading to different transport patterns could therefore be the reason for this misrepresentation.

To get a more concise overview for the sensitivity studies, the stations are roughly grouped into the regions shown in Figure 3.2 by calculating the average of the monthly median values among these stations. Additionally, the maximum and minimum of the medians is indicated by the shaded area.

The Central Arctic stations of Alert, Barrow, Ny-Ålesund, Zeppelin Station, and Villum are combined into Figure 3.2a. The maximum near-surface concentrations were observed in March, which is known to be the time of Arctic Haze. Figure 3.2b was created in the same way as Figure 3.2a, but for the IMPROVE stations Alaska. Here the highest median concentrations were observed in the summer with an average median of $50\,\mathrm{ng\,m^{-3}}$ in July. ECHAM-HAM instead calculates a similar pattern as for the Central Arctic stations with slightly lower concentrations.

3.2. Vertical distribution of BC

Even though it is logistically difficult and expensive to perform aircraft campaigns in the Arctic, enough of them have already been performed by the scientific community to allow for a comparison with at least two different campaigns per season. Unless otherwise specified, a profile was created for each instalment of a campaign, by sampling from the model along the part of the flight track that is located north of 60°N and then

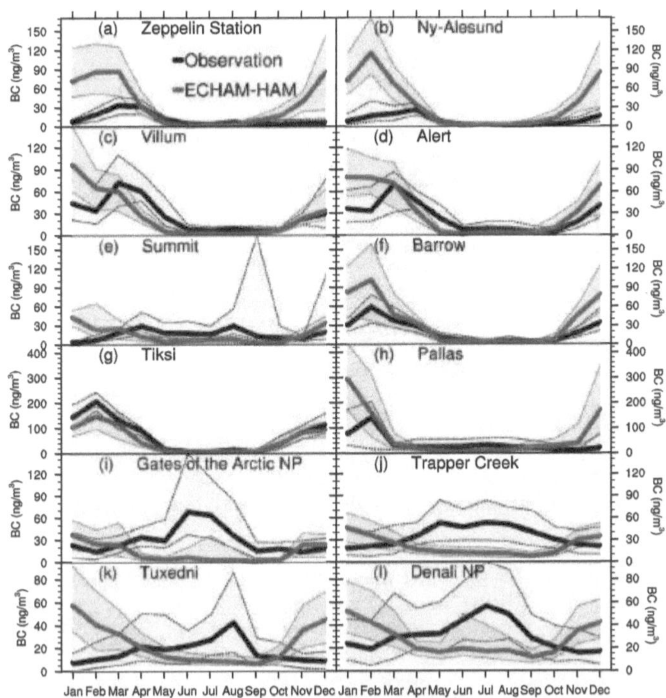

Figure 3.1.: Multi-year monthly statistics of the Arctic BC mass concentrations as observed and modelled in black and green, respectively. The median is indicated by the bold line, the quartiles are indicated by the dashed lines and shaded area. The geographical location and measured period of the observation sites are provided in Table 2.1. See also Figure 2.3 for the location in the Arctic.

calculating the mean profile. Then these mean profiles are averaged over the season including all campaigns available. Additionally, the maximum and minimum among the mean profiles is given for each pressure level.

The reason why the mean was chosen for the vertical profiles is the much lower sample size and coarse model resolution. A plume in the model can be less confined and has slightly higher concentrations everywhere instead of just in the small region where it would be observed, it is comparable to an spacial average over the gridbox. An averaging of both measurement and model therefore comes closer to comparing the same feature. The median potentially could show a high bias of the model that is only caused by the resolution, if multiple different data points are measured outside of the plume and few inside.

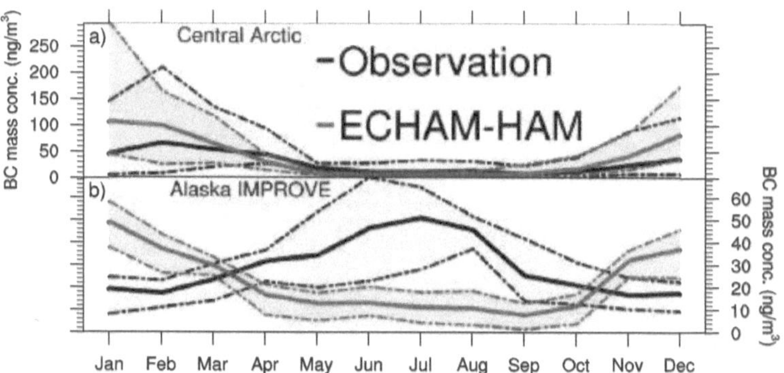

Figure 3.2.: Station data grouped by region. The bold line shows the average of the multi-year monthly medians of the Arctic BC mass concentrations as observed and modelled, in black and green, respectively. The maximum and minimum median concentrations in the regional groups are indicated by the dashed lines and shaded area.

In autumn, September-October-November (SON), two campaigns are available. The resulting profile from HIPPO-2 and ATom-3 are shown in Figure 3.3a. The measurements show small mean BC mass mixing ratios of less than $40\,\text{ng}\,\text{kg}^{-1}$, with the highest concentration near the surface and a local maximum of around $20\,\text{ng}\,\text{kg}^{-1}$ at $450\,\text{hPa}$. ECHAM-HAM produces similar mass mixing ratios, but the modelled BC values do not decline with altitude as quickly as in the observations, overestimating mass mixing ratios in the lower to middle free troposphere. Additionally, the model consistently shows increasing BC mass mixing ratios with height above $400\,\text{hPa}$, a feature that was not observed.

For the winter months, December-January-February (DJF), HIPPO-1 and ATom-2 are available. The average, minimum and maximum of the mean profiles can be seen in Figure 3.3b. BC mass mixing ratios are small in winter. The measured and computed average profiles both show a maximum in the near surface BC mass mixing ratio, they were very high for HIPPO-1 influencing the mean at the lower levels strongly. The measured and modelled profiles agree quite well. However, the model slightly overestimates the mass mixing ratios in the free troposphere. The feature of increasing amounts of BC with height above $400\,\text{hPa}$ can also be seen for DJF.

The spring, March-April-May (MAM), profile is very similar between model and measurements, as can be seen in Figure 3.3c. The highest observed mean profile is underestimated by the model, which is to be expected since the aircraft campaign (ARCTAS-spring) was actively measuring biomass burning plume BC mass mixing ratios, creating a high bias. The plume height is reproduced by ECHAM-HAM in this case. The average of the mean profiles agrees well up to a height of roughly $350\,\text{hPa}$, however, above

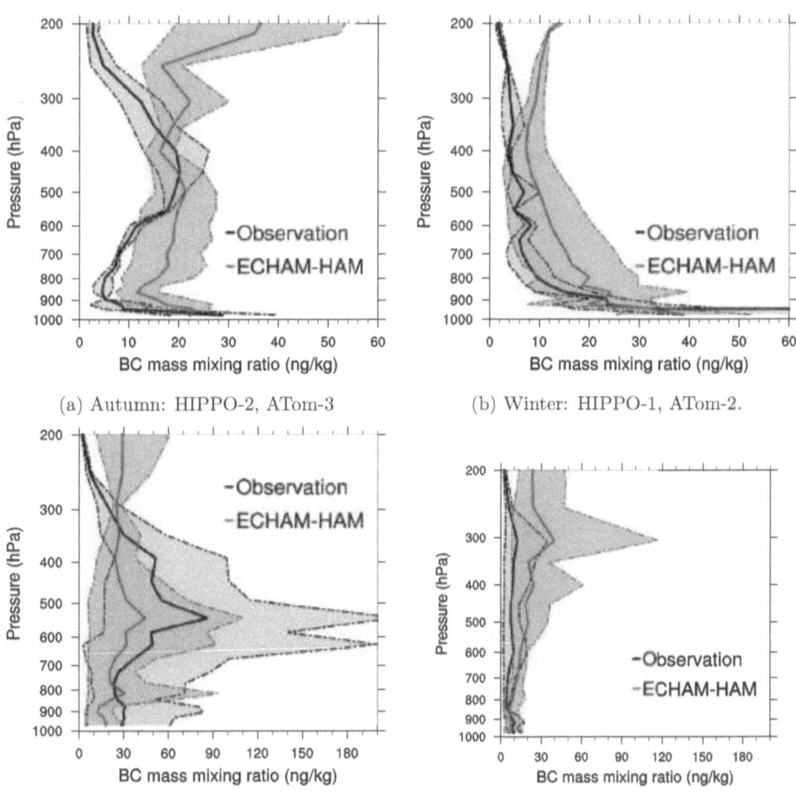

(a) Autumn: HIPPO-2, ATom-3

(b) Winter: HIPPO-1, ATom-2.

(c) Spring: ARCTAS-spring, ACLOUD, PA-MARCMiP, HIPPO-3, ATom-4.

(d) Summer: ARCTAS-summer, ACCESS, HIPPO-4, HIPPO-5, ATom-1.

Figure 3.3.: Vertical profiles of the BC mass mixing ratios in each season as observed and modelled during the campaigns in that season. The average of the mean profiles from the campaigns is indicated by the bold line for observations and the model in black and green, respectively. The dashed line/shaded area show the maximum and minimum from the mean profiles. See also Figure 2.3 for the geographic regions where the airborne observations were conducted.

this ECHAM-HAM again produces higher mass mixing ratios than could be observed.

In summer, June-July-August (JJA), both the measured and modelled profile (see Figure 3.3d) show a slight increase of BC mass mixing ratios with height. This is consistent with the transport theories of a lower atmospheric blocking during summer. There is a good fit between both profiles, but the increase with height is stronger in ECHAM-HAM. The number of campaigns in summer is higher than in autumn and winter, making this profile more robust. The importance of the strong overestimation by ECHAM-HAM in the peak at 300 hPa is related to a biomass burning event and should not be seen as a shortcoming of the aerosol microphysics model. It will be discussed in detail in connection to the sensitivity to emissions. Above 300 hPa, a decrease in BC concentrations is observed, which is not reproduced by the model.

ECHAM-HAM produces vertical BC profiles in the Arctic reasonably well. However, there is a clear problem of an overestimation in the upper troposphere/lower stratosphere that persists throughout the year.

Chapter 4.

Sensitivity studies

4.1. Emission related uncertainties

As discussed earlier, BC emissions are one of the largest sources of uncertainty for modelling the Arctic BC distributions and therefore radiative effects of BC. In order to investigate the uncertainty range in the Arctic BC distribution resulting from the uncertainty in emission, the sensitivity of ECHAM-HAM is examined. For this purpose four different state-of-the-art emission inventories are used. The setups are chosen in a way that they span a range of assumptions that are reasonable to use in aerosol climate model studies, depending on the focus of the study. When compared against each other they cover three main areas that can be the source of uncertainty:

- A regional refinement is used to show the importance of correctly capturing regionally important sources correctly. The original emission dataset covers Russia on a very fine resolution (0.1°) and has a special focus on the regionally important emission source of gas flaring. It is converted to the resolution of ECHAM-HAM to make it usable.

- Fixed emissions for the year 2000 are used to show the importance of using an emission dataset that is representing recent changes in the global economic situation and emission regulations.

- Monthly changing emissions that are fixed for the year 2000 show the impact that a use of daily biomass burning emissions that are tied to satellite retrievals has.

The configurations are described in detail in subsection 2.3.1.

BCRUS, which is named after the updated Russian emissions from K. Huang et al. (2015) is used as the reference run. The annual mean emissions are shown in Figure 2.5a. This setup is included as a state-of-the-art emission inventory with a high estimate of Russian gas flaring emissions, a source of pollution that has been discussed as possibly too low in commonly used emission datasets (Eckhardt et al. 2017). The resulting annual mean BC burden, that is the vertically integrated BC mass, is shown in Figure 4.1a. It features a rough pattern of a more polluted Eastern hemisphere, with the highest burdens in the main source regions of Europe, Russia and China. This pattern is also visible inside the Arctic, where the lowest burden is found over the Greenland ice sheet with less than $100\,\mu g\,m^{-2}$. North of 60°N the Eastern Hemisphere shows BC burdens of 200 to $800\,\mu g\,m^{-2}$, compared to 50 to $400\,\mu g\,m^{-2}$ in the Western Hemisphere. This separation is a result of the distribution of emissions, which are higher in the Eastern Hemisphere. *BCRUS* exhibits the highest Arctic (>60° N) annual multi-year area-weighted mean burden among the four setups with $254\,\mu g\,m^{-2}$. The inter-annual variability is $22\,\mu g\,m^{-2}$. The highest values north of 60°N are located in the Russian gas flaring region, at more than $560\,\mu g\,m^{-2}$ on average.

The annual average vertical profile in the Arctic north of 65°N for *BCRUS* is shown in green in Figure 4.2a for the year 2008. The mass mixing ratios are higher than for the other emission configurations and show a maximum at the surface caused mainly by the high Russian emissions. These Russian gas-flaring pollutants tend to stay near the surface because of the typical atmospheric stratification of the Arctic lower troposphere.

4.1.1. Regional refinement

The reference run *BCRUS* includes a regional refinement of BC over Russia, with a significantly higher spatial resolution in this area and more recent estimates of gas flaring pollution (as mentioned above) compared to the base anthropogenic emission inventory ECLIPSE. In order to quantify the impact of the regional refinement on the BC distribution, the run *BCRUS* is compared to the run *ECLIPSE*, which does not include the Russian BC emission refinement. The run ECLIPSE uses the same anthropogenic emissions of ECLIPSE v5a as run BCRUS, except in Russia, which explains the small difference in emissions between the two, which is shown in Figure 2.5c. This setup is used in combination with *BCRUS* to estimate this effect of Russian gas flaring emissions, while also providing the same recent and evolving emissions over the simulation time. This change in Russian emissions, while regionally improving the accuracy, does not impact the spatial distribution of BC outside of Russia and the Russian Arctic meaningfully, as can be seen in Figure 4.1c. The increase over the Barents Sea and Kara Sea, however, is higher than $25\,\mu g\,m^{-2}$. The annual mean area-weighted change in burden north of 60°N is $11\,\mu g\,m^{-2}$ less in *ECLIPSE* than in *BCRUS*. Because of the common stratification of the Arctic lower troposphere, this increase in burden is dominated by an increase in the lower tropospheric BC concentrations, as can be seen in Figure 4.2. The *ECLIPSE* run (in blue) produces on average lower values by $7\,ng\,kg^{-1}$. This feature is especially important for the near surface temperatures and therefore the Arctic Amplification, because BC warms the surrounding air when it absorbs solar radiation.

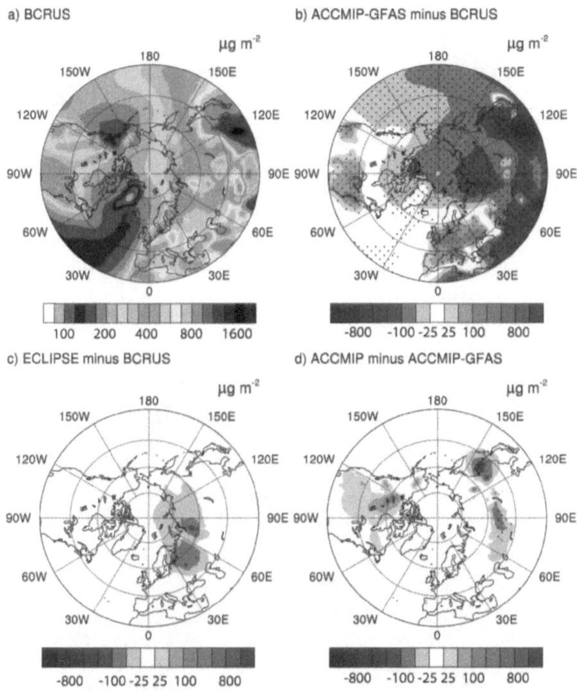

Figure 4.1.: Maps of annual mean BC burden for the years 2005-2015. (a) Absolute values are given for *BCRUS*. Difference between (b) the *ACCMIP-GFAS* and *BCRUS* results, (c) the *ECLIPSE* and *BCRUS* results and (d) between the *ACCMIP2000* and *ACCMIP-GFAS* results.

4.1.2. Recent economic changes

The run ACCMIP-GFAS uses the ACCMIP emissions for the year 2000, which are a useful and common choice for non-transient simulations of model inter-comparison studies. ACCMIP-GFAS is compared against BCRUS, which uses evolving anthropogenic emissions. In this way the uncertainty can be quantified that is introduced when recent economic changes and changes in government policies are not taken into account. Like the previous two setups, both runs here use the GFAS daily biomass burning emissions, which cover natural and human-caused fires. Therefore, the differences between *ACCMIP-GFAS* and *BCRUS* are solely due to the different setup in terms of anthropogenic emissions. Figure 2.5b shows how the annual mean emissions differ between *ACCMIP-GFAS* and *BCRUS*.

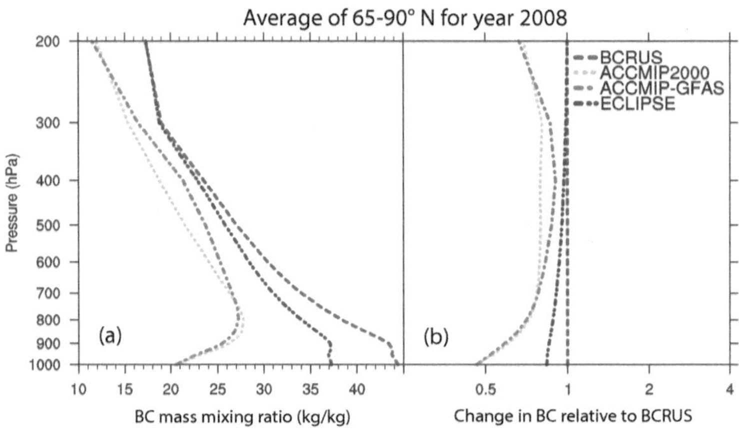

Figure 4.2.: Average Arctic (65-95° N) profile of (a) the BC mass mixing ratio and (b) the change in the BC mass mixing ratio relative to the base run *BCRUS*. See subsection 2.3.1 for a detailed definition of the different emission setups.

When using the fixed anthropogenic emissions in *ACCMIP-GFAS*, the atmospheric burden of BC is remarkably different in the source regions (see Figure 4.1b. BC burdens in *ACCMIP-GFAS* are considerably higher than in *BCRUS* over Europe, North America and Japan, where strict air quality legislation caused a reduction in BC emissions in recent years (up to over $200\,\mu g\,m^{-2}$ more). In contrast, they are much lower in China (regionally by more than $1200\,\mu g\,m^{-2}$ less), because of the massive growth of the Chinese economy since the year 2000. North of 60°N the *ACCMIP-GFAS* underestimates the BC burden by $63\,\mu g\,m^{-2}$ (25 %) compared to *BCRUS*. Over the Kara Sea, the underestimation exceeds $100\,\mu g\,m^{-2}$. This region has been identified as linked to the large-scale atmospheric circulation, by being particularly sensitive to a sea-ice loss (e.g., Petoukhov et al. 2010).

The Arctic BC mass mixing ratio vertical profile of *ACCMIP-GFAS* can be seen in red in Figure 4.2a, the difference in the profile relative to *BCRUS* in Figure 4.2b. The two runs differ strongly in the troposphere below $700\,hPa$ with *ACCMIP-GFAS* producing roughly half of the BC at the surface and above $300\,hPa$ with roughly one third less BC. The profile of *ACCMIP-GFAS* shows a different pattern with a minimum at the surface of slightly more than $20\,ng\,kg^{-1}$, which is only half of the BC mass mixing ratio as for *BCRUS*. *ACCMIP-GFAS* has much lower local sources, which is the reason for the shape of the vertical layering in combination with the typical stratification of the lower Arctic troposphere.

4.1.3. Temporal variability in wildfire events

The run ACCMIP2000 is the only run not using daily changing satellite observation based biomass burning emissions and is therefore also useful for studies using non-transient simulations. The differences between ACCMIP2000 and ACCMIP-GFAS show the uncertainty range introduced by not using realistic biomass burning emissions derived from satellite retrievals (see Figure 2.5d). The global BC emissions of ACCMIP2000 are lower by $64.5\,\mathrm{kt\,yr^{-1}}$ than in ACCMIP-GFAS. The composition of global aerosol and the BC burden are strongly influenced by biomass burning. The resulting change in the multi-year annual mean BC burden, displayed in Figure 4.1d, shows an overestimation caused by the fixed year 2000 fires over North America and an underestimation roughly along the southern Russian border from the Aral Sea to the Pacific Ocean. In the Arctic this results in a change in the area-weighted multi-year annual mean burden of only $11\,\mathrm{\mu g\,m^{-2}}$ (6 %), with lower values for ACCMIP2000. The averaged vertical profiles for the year 2008 in Figure 4.2 are comparable to the ones of ACCMIP-GFAS, with only slightly lower values in the free troposphere between $700\,\mathrm{hPa}$–$300\,\mathrm{hPa}$.

The spatio-temporal variability of biomass burning is very high, which leads to strong local influences that are not captured by this comparison of multi-year annual means, but will become apparent in the evaluation against measurements.

4.1.4. Comparison against ground observations

As done in section 3.1, the long time series of BC measurements are used to evaluate the modelled BC distribution and seasonal cycles. The main question is, whether any of the emission setups appear to be advantageous and which errors are introduced by the assumptions implied in them.

Instead of evaluating the seasonal cycle of the BC concentrations of a single station at a time, they are averaged for two regions. The average of the median concentrations is shown in Figures 4.3a and b by bold lines. The shaded area shows the maximum and minimum among the median concentrations in that ensemble. For the Central Arctic Figure 4.3a, ECHAM-HAM tends to overestimate the BC concentrations from November to February and slightly underestimates surface BC in April. The generally higher concentrations of BCRUS mean that the period of the overestimation reaches into March, while this is not the case for the other runs. However, the underestimation in April is also slightly lower.

All runs produce a very similar seasonal cycle, that is only scaled by a factor that seems to be related to the total BC emissions in the respective setup.

The IMPROVE stations in Alaska are shown in Figure 4.3b. Here the problem with the shift in the seasonal cycle by about six months that was discussed in section 3.1 is present for almost all runs. The only noticeably different run is *ACCMIP2000*, depicted in orange, which produces a summer maximum with roughly the correct concentrations. Since this run only differs from *ACCMIP-GFAS* by the biomass burning emissions, these must be the reason for the seemingly better performance. In some aspects this makes sense since wildfires are the most likely local source of BC in summer. However, the biomass burning emissions in *ACCMIP2000* are fixed to the year 2000 and can therefore not be more realistic, producing these emissions exactly in the same time period and location with the same strength every year. It is however possible that regular, very small fires, that are not detected by the satellite retrieval used in GFAS, cause this maximum, while the overestimation in winter is just the typical behaviour that ECHAM-HAM shows for the other Arctic stations.

Figure 4.3 shows a separate station, Tiksi (that is also included in the average of Figure 4.3a), because of its unique location in Russia, where no other station data were available. The central thick lines in Figure 4.3 therefore show the median of the multi-year time series, with the dotted lines showing the interquartile range. At this station, close to the source of gas flaring in the Russian Arctic, *BCRUS* (in green; see also subsection 4.1.1) is clearly the best at reproducing the observed seasonal cycle, while the other runs underestimate the concentrations. Since correct emissions become more important, with closer proximity to the source, the microphysics become more important the longer the transport pathway is, this station shows that *BCRUS* could indeed be the best setup in terms of emissions.

Reproducing plumes of enhanced BC concentrations at the correct time and in the correct strength is very challenging for a global model at a coarse horizontal resolution, but still a desirable capability. To evaluate how the different emission setups influence modelled BC level in the Arctic, the Pearson correlation coefficients between the collocated data of measured and modelled BC mass concentrations for all available stations in the Arctic were calculated. They are shown as colour-coded circle segments in Figure 4.4 and in Table 4.1. The top right segment of each circle shows the correlation coefficient between the *BCRUS* model run and the measurements. Following in clockwise order are the runs *ACCMIP2000*, *ACCMIP-GFAS*, and *ECLIPSE*.

The circle for Summit is not filled, because the Pearson correlation coefficient is negative for all runs, with e.g. -0.06 for *BCRUS*. The negative correlation is caused by the shift in the seasonal cycle by roughly six months (see Figure 3.1e). All other stations show a positive correlation coefficient.

Even though the seasonal cycle for the IMPROVE stations in Alaska was not well reproduced by ECHAM-HAM, two of them show good correlation coefficients for *BCRUS*, with 0.72 and 0.94 for Denali NP, and Gates of the Arctic NP, respectively. Two stations show a correlation in the range of most other Arctic stations, with 0.44, and 0.55, for Tuxedni and Trapper Creek, respectively. For Simeonof, however, all runs show almost no correlation with roughly 0.1 for all runs. *ACCMIP2000* performs the worst of all the setups with correlation coefficients of 0.31, 0.20, 0.02, and 0.14 for Denali NP, Gates of the Arctic NP, Tuxedni, and Trapper creek, respectively. The other runs

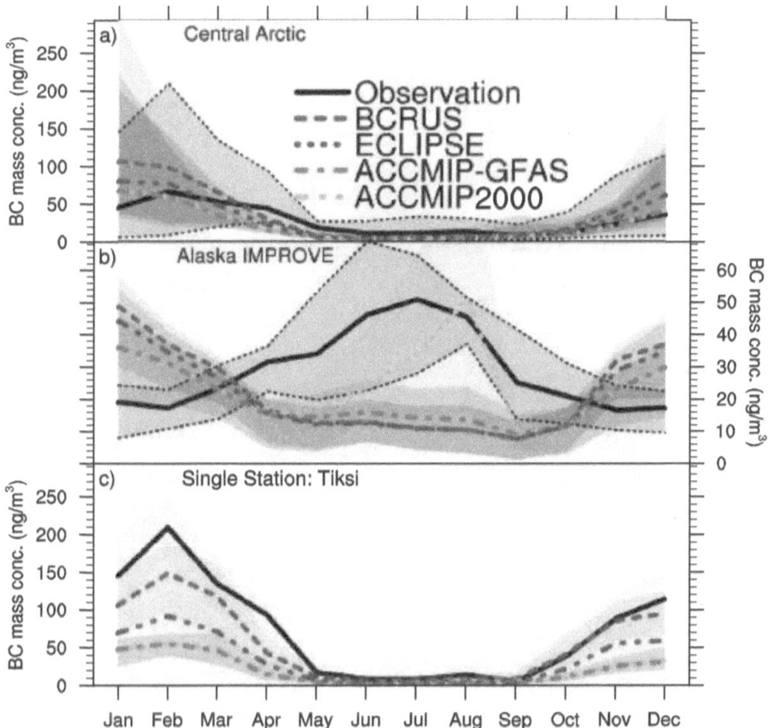

Figure 4.3.: Station data grouped by region. The bold line shows the average of the multi-year monthly medians of the Arctic BC mass concentrations as observed and modelled, in black and green, respectively. The maximum and minimum median concentrations in the regional groups are indicated by the dashed lines and shaded area.

Table 4.1.: Pearson correlation coefficients between observations and measurements for all available Arctic stations..

Station	BCRUS	ECLIPSE	ACCMIP-GFAS	ACCMIP2000
Alert	0.50	0.49	0.48	0.48
Oulanka	0.23	0.28	0.29	0.27
Barrow	0.74	0.73	0.68	0.68
Ny-Ålesund	0.53	0.52	0.46	0.46
Villum	0.62	0.62	0.63	0.64
Tiksi	0.56	0.61	0.72	0.37
Pallas	0.50	0.46	0.51	0.54
Summit	-0.02	-0.02	-0.03	-0.04
Zeppelin Station	0.32	0.29	0.28	0.31
Gates of the Arctic NP	0.94	0.95	0.94	0.20
Trapper Creek	0.55	0.56	0.56	0.14
Tuxedni	0.44	0.48	0.47	0.02
Simeonof	0.09	0.10	0.12	0.14
Denali NP	0.72	0.74	0.75	0.31

do not differ strongly from each other. In contrast to the seasonal cycle the runs with GFAS perform much better in this metric. This again indicates that Alaska is often affected by wildfires and that a satellite observation based representation of fire emission on a daily basis is key. The combination of the two leads to the conclusion that the runs with GFAS compute biomass burning plumes at the right time, but not in the correct intensity. Without GFAS or a similar biomass burning inventory this is not possible.

The stations in the European Arctic show relatively low correlation coefficients for *BCRUS* with 0.50, 0.53 and 0.32 for Pallas, Ny-Ålesund, and Zeppelin Station, respectively. For the fourth station in the European Arctic, Oulanka, the correlation coefficient is less than 0.3 for all runs. However there were only 3 months of measurements available for this station. The correlation coefficients for all four stations do not differ much between the runs, showing that these sites are much less affected by relatively local biomass burning events, but by different transport or blocking situations, mainly from the European and Russian source regions. These correlation coefficients therefore show more about the accuracy of the long range transport, than about the correct timing of emissions.

Utqui'gvik (Barrow), Alert, Villum Research Station and Tiksi are all located north of a big landmass. Since concentrations likely change sharply depending on whether the wind comes from land or the Arctic Ocean, their correlation coefficients are good, with 0.74, 0.50, 0.62, and 0.56 for *BCRUS*, respectively. Only at Tiksi, one run performs considerably worse than the others, which is *ACCMIP2000* with a correlation coefficient of 0.37. Tiksi therefore also seems to be affected by BC from biomass burning events. Interestingly, the highest correlation coefficient is not found for *BCRUS* with the regional improvements, but for *ACCMIP-GFAS*, with 0.72. The *ECLIPSE* run also produces a higher correlation coefficient with 0.61. This seems to be contradicting the improved seasonal cycle with *BCRUS* of Tiksi (see Figure 4.3c). This behaviour could be explained by the different distribution and spatial resolution of sources between *BCRUS* and *ECLIPSE*, that could cause this difference in combination with the local orography being misrepresented in the coarse model resolution.

4.1.5. Comparison with aircraft campaigns

To assess how the different emissions datasets impact the capabilities of ECHAM-HAM to reproduce observed profiles, the four different runs are compared here. Instead of the averaged mean vertical profiles in section 3.2, here all campaigns are shown separately. There are two reasons for this. Only the run *BCRUS* includes the years 2016–2018, limiting the number of available campaigns in autumn and winter to only one campaign per season. More importantly, it is of interest whether the different setups are favourable for different reasons.

Autumn

The single profile for autumn in Figure 4.5 was taken during the HIPPO-2 campaign in November 2009. Flights were located over the Pacific part of the Arctic, as indicated by the blue box in Figure 2.3. Observations and the model show a maximum in the BC mass mixing ratio at the surface, which is underestimated by the model. This underestimation is smaller for *BCRUS* than in any other run with $28\,\mathrm{ng\,kg^{-1}}$ compared to the measured $39\,\mathrm{ng\,kg^{-1}}$. Both runs that use ACCMIP show a more pronounced underestimation, with roughly the same mass mixing ratios of $20\,\mathrm{ng\,kg^{-1}}$. In the free troposphere, both the model and measurements show a plume with increased BC mass mixing ratios. All model runs position it too low by about $200\,\mathrm{hPa}$. However, the maximum of the *ECLIPSE* and *BCRUS* runs are roughly the same as the observed maximum of $26\,\mathrm{ng\,kg^{-1}}$. The *ACCMIP* runs produce a little lower concentrations in this plume with $20\,\mathrm{ng\,kg^{-1}}$. The overestimation above $300\,\mathrm{hPa}$ that was discussed in section 3.2 is more pronounced in *BCRUS* and *ECLIPSE*, as in most Arctic BC profiles, see Figure 4.2.

Winter

The only available campaign for winter is from the first instalment of the HIPPO campaign in January 2009. As for HIPPO-2, the flights were located over the Pacific Arctic (see blue box in Figure 2.3). The resulting profile of BC mass mixing ratios can be seen in Figure 4.6. The profile only starts at $950\,\mathrm{hPa}$, because some measurements below that level were unrealistically high leading to an average mass mixing ratio of more

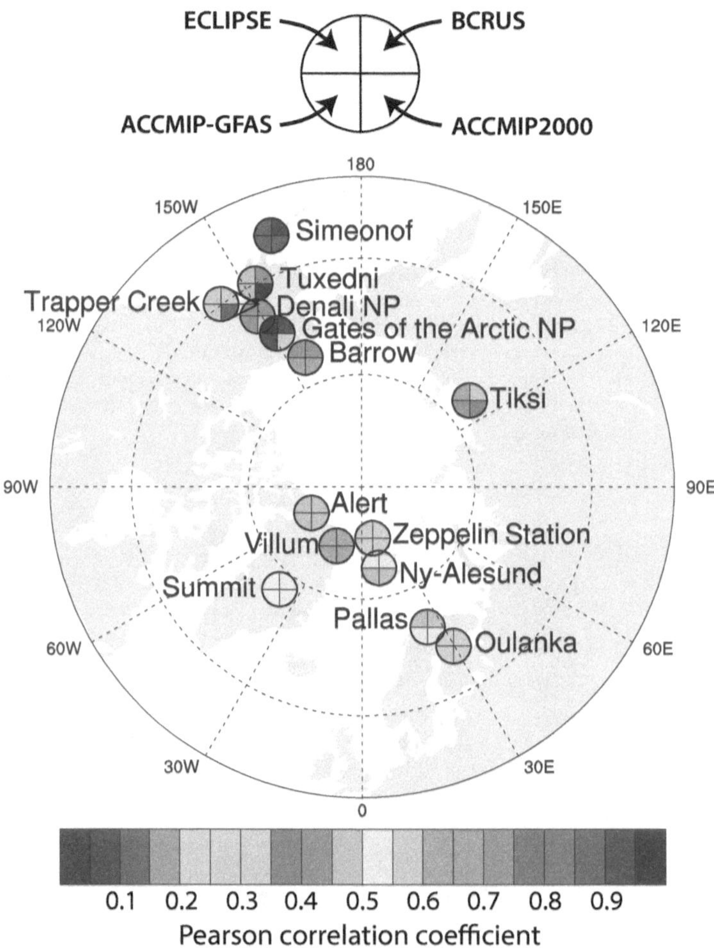

Figure 4.4.: Circles show Arctic sites with near-surface BC mass concentration measurements. The colours of the circle segments indicate the Pearson correlation coefficient between observations and the different emission setups. Correlation coefficients close to zero are not coloured. The top right circle segment represents the *BCRUS* run. *ACCMIP2000*, *ACCMIP-GFAS*, and *ECLIPSE* follow in clockwise order. The circle and label of Zeppelin station have been shifted to the north for better visibility. Trapper Creek's label and circle are shifted to the south-east.

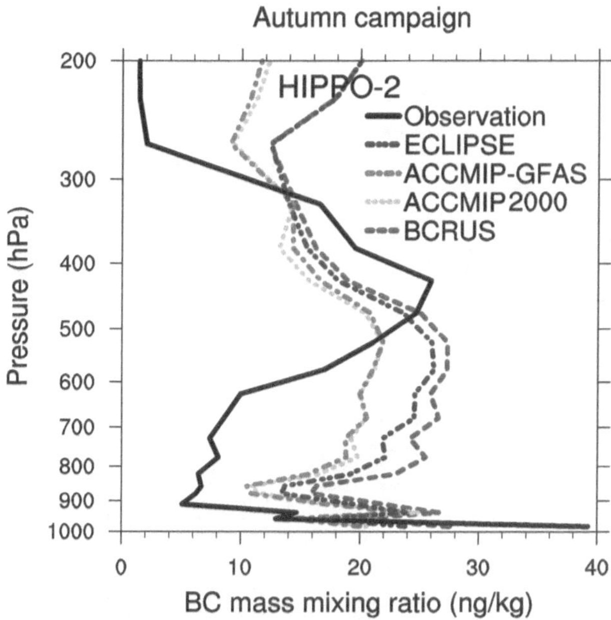

Figure 4.5.: Mean vertical profile of the BC mass mixing ratios in each campaign available in autumn between 2005 and 2015, as observed and modelled during the campaigns. Observations in black, the runs *BCRUS*, *ACCMIP2000*, *ACCMIP-GFAS* and *ECLIPSE* in green, orange, red, and blue, respectively.

than $450 \, \text{ng} \, \text{kg}^{-1}$. Above this, both the model and observations show a similar profile of declining mass mixing ratios with increasing height. The model overestimates the measured profile by about a factor of two. The runs *BCRUS* and *ECLIPSE* show roughly the same profile above 700 hPa, and a stronger overestimation than *ACCMIP2000* and *ACCMIP-GFAS*. *BCRUS* and *ECLIPSE* have the same emissions for Asia, that are higher than the ACCMIP emissions. The *BCRUS* and *ECLIPSE* emissions differ in Russia. The ACCMIP emissions are higher in North America. It is therefore likely that the overestimation in this profile is caused by an overestimation of the long-range transport from the non-Russian part of Asia.

Spring

The profiles for the two campaigns in spring can be seen in Figure 4.7. The first one (Figure 4.7a) shows observations and model results for the ARCTAS spring campaign in April 2008. It is one of the only cases where the profile of measured BC mass mixing

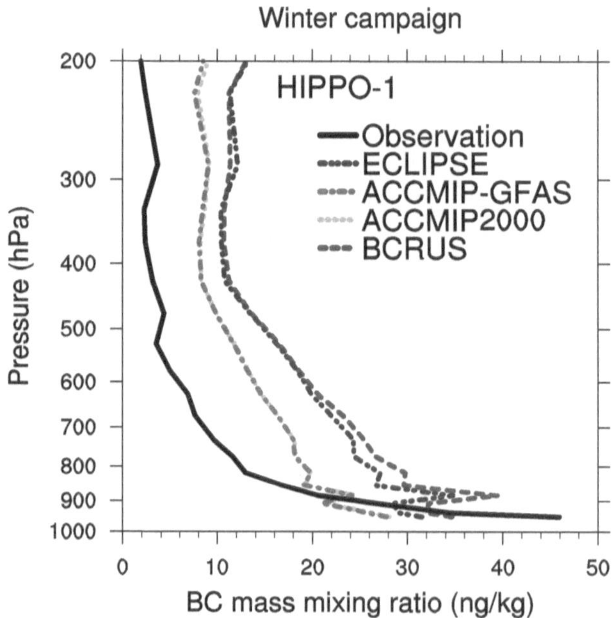

Figure 4.6.: As in Figure 4.5 but for winter.

ratios is underestimated by ECHAM-HAM almost throughout the profile, independent of the emission setup. This is largely caused by the design of the ARCTAS campaigns. They both intendted find and measure biomass burning plumes in the Arctic. This leads to the highest observed mass mixing ratios among the profiles available, with values of up to $210\,\mathrm{ng\,kg^{-1}}$ in the averaged profile. The plume is well produced in the model for all runs except *ACCMIP2000*, which is an indicator that the plume in the model was indeed only reproducible because of the daily changing biomass burning emissions of GFAS. The underestimation by about a factor of two to three is at least in parts caused by the coarse model resolution that distributes the high concentration over the whole grid. Therefore, it cannot be expected to reproduce mass mixing ratios as high as the ones that were measured in the actively targeted plume.

The BC mass mixing ratio profiles for HIPPO-3 are shown in Figure 4.7b. It took place in March to April 2010. The observed profile looks comparable to the one of ARCTAS-spring, the mass mixing ratios are much smaller with the maximum in the averaged profile in $520\,\mathrm{hPa}$ with only $38\,\mathrm{ng\,kg^{-1}}$. Near the surface *ECLIPSE* and *BCRUS* are able to reproduce the observed mass mixing ratios better than *ACCMIP2000* and *ACCMIP-GFAS*. Up to $700\,\mathrm{hPa}$ ECHAM-HAM produces similar mass mixing ratios as were observed. From there up to $400\,\mathrm{hPa}$ all runs produce too low values. Further up

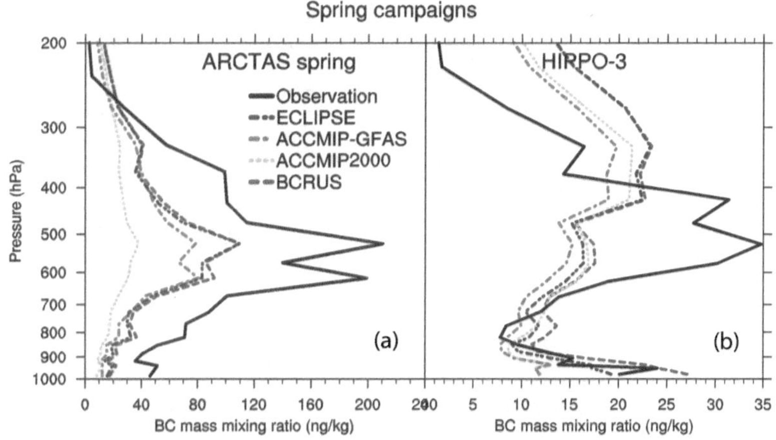

Figure 4.7.: As in Figure 4.5 but for spring.

the modelled concentrations are higher. *BCRUS* and *ECLIPSE* are further away from the observations at this height than *ACCMIP2000* and *ACCMIP-GFAS*.

Summer

With four campaigns, summer is the season with the most campaigns available. The vertical profile for the campaigns ACCESS, HIPPO-4, HIPPO-5, and ARCTAS-summer are shown in Figure 4.8. Observed values are generally low with under $20\,\mathrm{ng\,kg^{-1}}$ for all mean campaign profiles except the plume observed in ARCTAS-summer. *ACCMIP2000* shows profiles for all campaigns that are remarkably different from all other runs, showing the importance of wildfires for the Arctic BC concentrations especially in summer.

The BC mass mixing ratio profile of ACCESS, observed in June 2012, is shown in Figure 4.8a. It shows an overestimation by the model of about $10\,\mathrm{ng\,kg^{-1}}$ above $850\,\mathrm{hPa}$. Since *ACCMIP2000* differs from *ACCMIP-GFAS* which agrees well with *ECLIPSE* and *BCRUS*, this overestimation is likely caused by BC from biomass burning emissions. These runs show a peak concentration of $24\,\mathrm{ng\,kg^{-1}}$ at about $400\,\mathrm{hPa}$. At this height the observations also show a maximum, which is less pronounced with a peak mass mixing ratio of only $13\,\mathrm{ng\,kg^{-1}}$.

For HIPPO-4, which took place in June to July 2011, the measured and modelled profiles are presented in Figure 4.5b. The level of BC mass mixing ratio is well reproduced, but the vertical distribution in ECHAM-HAM is different. All runs except *ACCMIP2000* produce mass mixing ratios that are comparable to the observations.

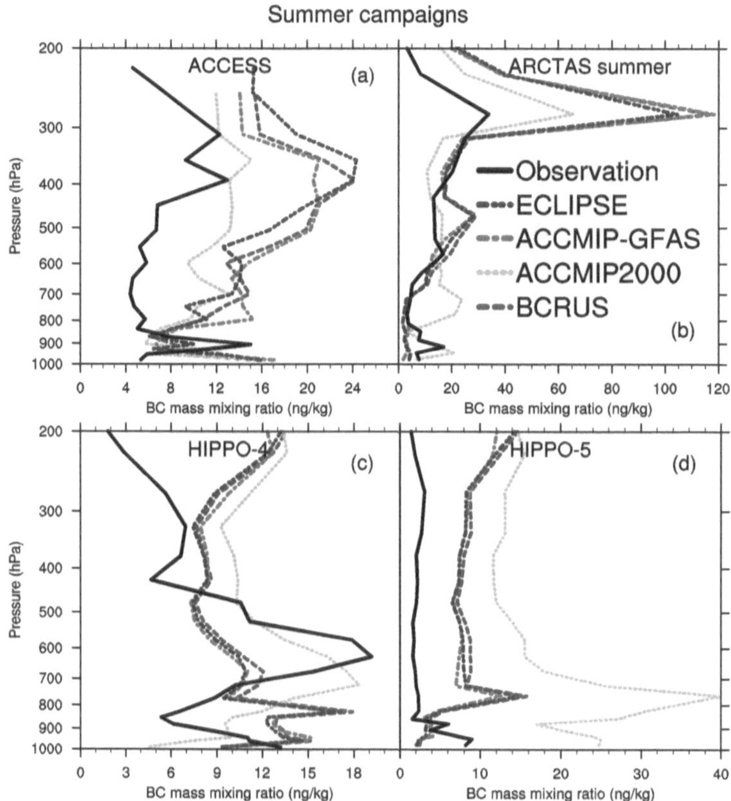

Figure 4.8.: As in Figure 4.5 but for summer.

However, ECHAM-HAM computes increasing concentrations with increasing altitude, to a maximum of $18\,\mathrm{ng\,kg^{-1}}$ at $820\,\mathrm{hPa}$, while the observations show a layer with cleaner air with only $5\,\mathrm{ng\,kg^{-1}}$ instead. In the observations a polluted layer with similar peak mass mixing ratios can be found in $620\,\mathrm{hPa}$ instead. The common overestimation above $300\,\mathrm{hPa}$ can be found for this campaign as well.

The HIPPO-5 measurements were taken in August to September 2011. The corresponding BC mass mixing ratio profiles are shown in Figure 4.8c. They show extremely low observed BC mass mixing ratio values that are only slightly enhanced at the surface. The modelled values are higher, with *ACCMIP2000* performing the worst. None of the runs shows the near surface maximum that is present in the observations. The runs using GFAS emissions are showing near constant values of around $10\,\mathrm{ng\,kg^{-1}}$ throughout the profile with an increase above 280 hPa. In contrast to this, the observed profile is close to constant throughout the atmosphere with around $2\,\text{ng}\,\text{kg}^{-1}$ and a slight decline above 280 hPa instead.

The profiles for the fourth campaign in summer are shown in Figure 4.8d. ARCTAS-summer took place in June and July 2008. Like its sister campaign, it shows higher BC mass mixing ratios than the other campaigns in the season, because biomass burning plumes were specifically sampled. In contrast to ARCTAS-spring (Figure 4.7a), this time the modelled mass mixing ratios are higher, which is surprising given the issue of the coarse model resolution biasing the model towards lower mass mixing ratios. However, Matsui et al. (2011) have shown that the plume measured here has only slightly enhanced BC concentrations, but a more strongly enhanced CO (which is an additional indicator for fire plumes) concentrations at the altitude where ECHAM-HAM places the BC maximum. The observed plume was found to be subject to a precipitation event near the source region. This precipitation event, however, did not occur in ECHAM-HAM. This shows a dependence on factors in the model that lie outside of the aerosol microphysics or emissions and thus, outside of the scope of this work.

4.2. Removal and ageing related uncertainties

Removal and ageing are the other big sources of uncertainty for modelling the BC transport to the Arctic and the vertical distribution of BC in the Arctic. A set of sensitivity studies shows the impact of the ageing rate, that is controlled via the number of SU monolayers needed to age BC, the SU emission strength which directly controls the amount of SU available to age BC, and the in-cloud scavenging efficiency, which is one of the most important processes removing BC from the atmosphere. These are the most likely to strongly influence the efficiency of the long-range transport of BC to the Arctic. The strength in which they are varied covers a physically reasonable range of assumptions. The technical details can be found in subsection 2.3.2. The runs in this section have a later starting date from the runs in earlier sections with January 2007, for more detail see Table 2.3.

4.2.1. Single parameter sensitivity

In the first part of this sensitivity study, parameters possibly influencing the wet removal of BC are modified separately. To gain additional information about the impact on the transport patterns BC tracers tagged with source regions are used. The tagged tracers are described in detail in section 2.1.1. The runs are shorter with only two years, which is sufficient for this sensitivity study, but also necessary, as the runs with tagged tracers are computationally more expensive. It should be noted here that a bias in the tagging method slightly increases the average BC amount in the Arctic. However, this discrepancy is considered negligible, as only tagged and non-tagged model results are compared to each other, respectively.

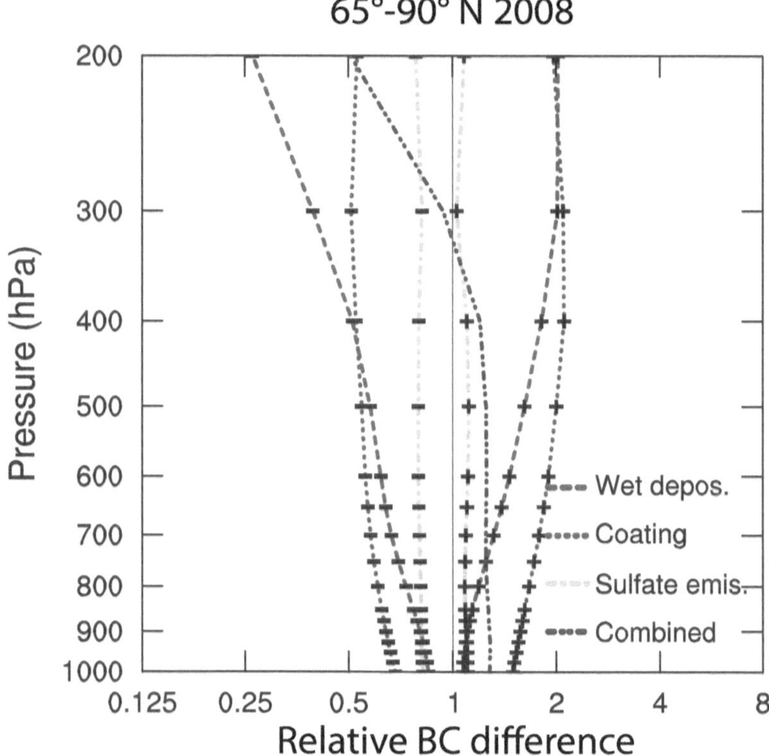

Figure 4.9.: Changes in the vertical Arctic BC mass concentrations relative to the base run *BCRUS-BCTAG*. Each pair of lines in the same colour (and with the same dash pattern) results from a different parameter.

In-cloud scavenging

The in-cloud scavenging is the collection of aerosol particles by cloud droplets or ice crystals, either through the aerosol particles acting as CCN or INP, or by collision between the aerosol particle and a hydrometeor. The parametrisations described in section 2.1.1 generate a number of aerosol particles that are partly removed by wet deposition in case of precipitation. The in-cloud wet scavenging efficiency is scaled up by a factor $\alpha_{ic} = 2.0$ and in another experiment down by $\alpha_{ic} = 0.5$ as described in Equation 2.6.

To evaluate the sensitivity it is useful to compare the vertical profiles of BC to those of a base run, as in Figure 4.9. The differences in the annual Arctic average (north of 65° N) BC profiles for year 2008 are shown in brown for this set of runs relative to the base run. The run with an increased wet deposition efficiency is expected to produce lower atmospheric concentrations of BC and is therefore marked with a minus sign in Figure 4.9. Near the surface the sensitivity is relatively low, but strongly increases with altitude. This is especially prominent in the run with increased wet deposition efficiency, causing decreases of more than 50 % above 400 hPa, reaching a reduction of almost 25 % of the BC mass mixing ratio compared to the unperturbed run at 200 hPa.

Since the complementary run with a decreased in-cloud scavenging is expected to produce increased concentrations, it is marked with plus signs in Figure 4.9. The amount of increase in BC concentrations for this run, while still showing a high sensitivity, is a little lower than in the run with enhanced in-cloud scavenging. The difference relative to the base run is also increasing with height, but in a way that concentrations roughly double compared to the base run.

The absolute changes in the average BC profile in the Arctic average (north of 65° N) of these two setups can be found in Figure 4.10. The contributions of the BC tracers that were tagged by source region are indicated by different colours.

For a higher in-cloud scavenging the concentrations decrease for all source regions, as shown in Figure 4.10a. The strong decrease in the BC mass mixing ratio above 400 hPa is largely caused by a decrease in Chinese BC and BC from the rest of the world. See Figure 2.1 for the definition of the source regions. At 200 hPa the reduction of $15\,\mathrm{ng\,kg^{-1}}$ is almost exclusively caused by the reduction in the transport from these regions. The BC does not reach the necessary altitude, as it is washed out before by precipitation, removing large parts of these very long-travelled pollution plumes. The transport from North America, Europe and Russia is also strongly affected. The concentrations in the middle troposphere are decreased most strongly. Near the surface the decrease is the smallest, with a reduction of only slightly more than $7\,\mathrm{ng\,kg^{-1}}$.

The experiment with lower in-cloud scavenging almost mirrors the experiment with higher in-cloud scavenging, with the biggest difference between perturbed and base run at 300 hPa. This is again largely caused by differences in the transport from the rest of the world and China, which was tagged separately as one of the regions with the highest BC emissions even though it is very far away from the Arctic. Interestingly, the absolute effect of the reduction is larger at high altitudes than in the run with the increased in-cloud scavenging, but lower near the surface. The lower in-cloud scavenging experiment even exhibits a small decrease in near surface BC mass mixing ratios of BC with an Arctic origin, while it increases above. A possible reason for this could be that less of the emitted BC gets mixed back down by precipitation that evaporates before it reaches the surface.

This high sensitivity to the in-cloud scavenging is not surprising and has been discussed before in other studies with ECHAM-HAM (e.g., Hoose et al. 2008; Croft et al. 2010; Watson-Parris et al. 2019). In another study which focusses on Arctic aerosol, Bourgeois et al. (2011) find an underestimation in the long-range transport, which is the opposite of what is found in this study. This underestimation is reproducible by using ECHAM-HAM with the older in-cloud scavenging parametrisation with prescribed scavenging fractions (as described in Stier et al. 2005), whereas here the diagnostic scheme of Croft et al. (2010) is used, because it produced better results in a comparison of preliminary test runs. To solve this underestimation, Bourgeois et al. (2011) suggest a lower in-cloud scavenging especially in mixed-phase clouds to increase the long-range transport to the Arctic. This leads to the increase in Arctic BC mass mixing ratios, as intended by them. See also section 2.1.1 for details on the scheme used here.

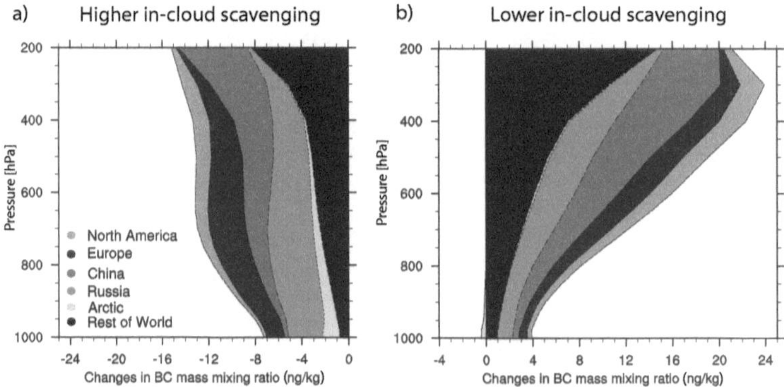

Figure 4.10.: Absolute changes in the vertical Arctic (north of 65° N) BC mass concentrations tagged by the different source regions, caused by changes in the in-cloud scavenging, compared to the base run *BCRUS-BCTAG* for the year 2008.

Slower ageing

The in-cloud scavenging efficiency of an aerosol particle type that is emitted as hydrophilic, depends on the rate at which it ages from hydrophilic to hydrophobic by collecting SU. The SU is assumed to form layers on the previously hydrophobic particles. The number of SU monolayers needed, and therefore the SU mass, is however uncertain. Different values can be found in the literature for different aerosol models. The base value of one needed monolayer in ECHAM-HAM is changed in this sensitivity study.

The decrease in the SU mass needed by $\alpha_{ml} = 0.3$ is expected to produce lower BC mass mixing ratios in the Arctic. The purple line in Figure 4.9 corresponding to this run is therefore marked with a minus sign. The shown sensitivity is much higher near the surface than for the experiment with increased in-cloud scavenging, with a decrease of the BC mass mixing ratio to about 70 % of the value of the reference run. This reduction means that an amount of SU that is not enough to cover the whole surface of an aerosol particle with a SU monolayer can already be enough to consider this particle hydrophilic. It is also much less altitude dependent reaching a reduction to roughly 50 % at 600 hPa, above which the difference stays almost constant.

The increase in BC concentrations caused by an increase in the required amount of SU needed to age a BC particle by $\alpha_{ml} = 5.0$ is shown by the purple curve marked with plus signs. This increase in needed SU means that a relatively thick layer of SU (five monolayers) is needed before an aerosol particle is considered hydrophilic. The expected increase roughly mirrors the decrease observed in the sister experiment. Near the surface it increases by about 70 %. At 600 hPa it reaches roughly double the mass mixing ratios of the base run. The maximum increase is reached at 400 hPa.

Figure 4.11 shows the absolute change compared to the base run in the BC mass mixing ratio by source region. The change in the BC mass mixing ratio profile for the experiment with a lower number of monolayers required, meaning faster ageing, is shown in Figure 4.11a. The effect on the transport from Europe and Russia is larger than in the in-cloud scavenging experiment. The strongest decrease is, therefore, found in lower altitudes, with a maximum of roughly $17\,\mathrm{ng\,kg^{-1}}$ at 900 hPa. While the sensitivity is still high, the affected transport pathways change in their relative importance.

The run with a slower ageing, caused by more required SU monolayers also shows this shift, in which transport is affected the strongest, as can be seen in Figure 4.11b. The maximum of the increase shifts a little higher to 600 hPa. This change with an increase in the BC mass mixing ratio of $30\,\mathrm{ng\,kg^{-1}}$ is the strongest change among the sensitivity experiments.

Sulphate abundance

Since SU is required by the parametrisation to age BC from hydrophobic to hydrophilic, as described earlier, the availability of SU should influence the wet deposition as well. This is done here by multiplying the emissions of SU by $\alpha_{SU} = 2.0$. The run with increased SU emissions is expected to produce lower BC burdens, since a higher abundance of SU will let BC age more quickly, allowing for more efficient wet removal. The

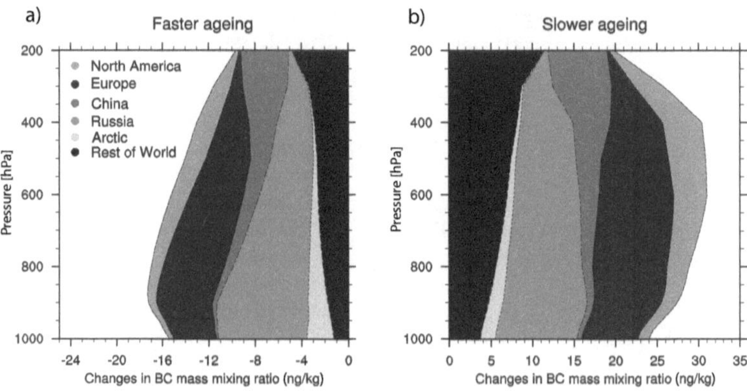

Figure 4.11.: As Figure 4.10, but for differences caused by changes in the aerosol ageing rate.

resulting change in the Arctic vertical BC mass mixing ratio profile relative to a base run is shown by the pink line marked with minus signs in Figure 4.9. The sensitivity to this parameter appears to be low, almost uniformly throughout the vertical profile at a value of about 80 % of the reference run. This relatively straight profile is interesting since a profile more similar to the run with changed coating thickness could be expected since both influence the same process. However, comparing to the run with $\alpha_{SU} = 0.5$, shown by the pink line marked with plus signs in Figure 4.9, reveals that the halving of the SU abundance has a smaller effect than the doubling.

The absolute change of the tagged BC tracers for the run with halved SU emissions can be found in Figure 4.12. Higher SU emissions affect the profile in a similar way as a faster ageing, the effect is however smaller with the parameters chosen here, see Figure 4.12. Their effect is so similar, because with a lower amount needed the available SU coats more BC and other particles close to the source regions, removing them from the atmosphere before they can be transported to the Arctic. The same happens with higher SU emissions, where the higher amount of available SU is sufficient for more particles. The effects on other parameters like total aerosol forcing will react in different ways. The maximum reduction of BC is at the same height.

Combination

In the evaluation of ECHAM-HAM (section 3.2), the model shows too high BC mass mixing ratios above roughly 300 hPa. However, at lower altitudes the model agrees well with observations. The strong reduction displayed by the experiment with the higher in-cloud scavenging makes it likely that changing this parameter would remove this

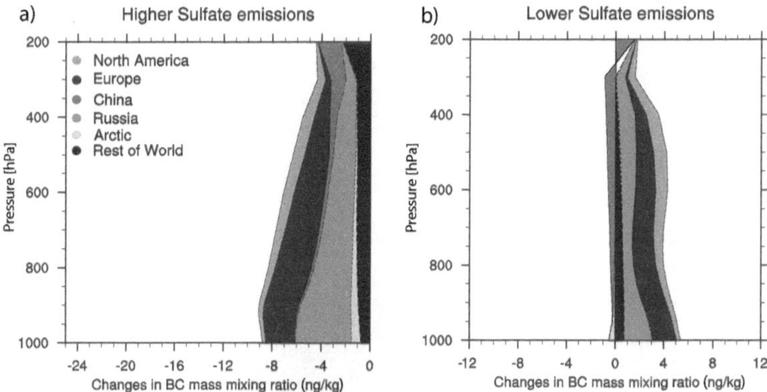

Figure 4.12.: As Figure 4.10, but for differences caused by changes in the SU emission flux.

issue. Since the sensitivity at other altitudes was also high, it would however lead to an underestimation there.

ECHAM-HAM also exhibits a high sensitivity to changes in the amount of SU needed to age BC, or in other words the ageing rate. It does however have a lower impact above 300 hPa. The number of required monolayers of SU needed to age an aerosol particle from hydrophobic to hydrophilic is very uncertain, making this change easily justifiable. A combination of a change in both of the two parameters seems to be, therefore, a promising possibility to improve the capability of ECHAM-HAM to reproduce the observed Arctic BC profiles.

Therefore, another sensitivity experiment was performed with $\alpha_{ml} = 5$ and $\alpha_{ic} = 2$. The resulting change in the vertical profile compared to the base run is shown in Figure 4.9 in blue. It exhibits a slight increase in the BC concentrations up to a level of 400 hPa, above which a reduction can be found. This is almost the desired behaviour, in order to improve the agreement between model results and aircraft measurements in the mid and upper troposphere. Still, this improvement comes with an increase in the middle and lower tropospheric BC concentrations, that will lead to an overestimation.

As clearly shown in Figure 4.13, the increase at the lower altitudes is considerable compared to the upper level reduction in absolute values. The BC transported from

China interestingly decreases throughout all altitude levels, indicating that the in-cloud scavenging is the dominating factor for this long-range transport pathway. The increase in BC mass below is roughly equally dominated by European and Russian BC. They show an increase throughout the profile. The increase in in-cloud scavenging seems not to stop BC from these regions reaching these high altitudes in the same way as it does for China. A possible explanation could be the higher amount of available SU in China, compared to Europe and Russia, still allowing the BC to age before it gets into contact with a cloud, even under the more prohibitive ageing threshold in this run, whereas in Europe and Russia this is not the case.

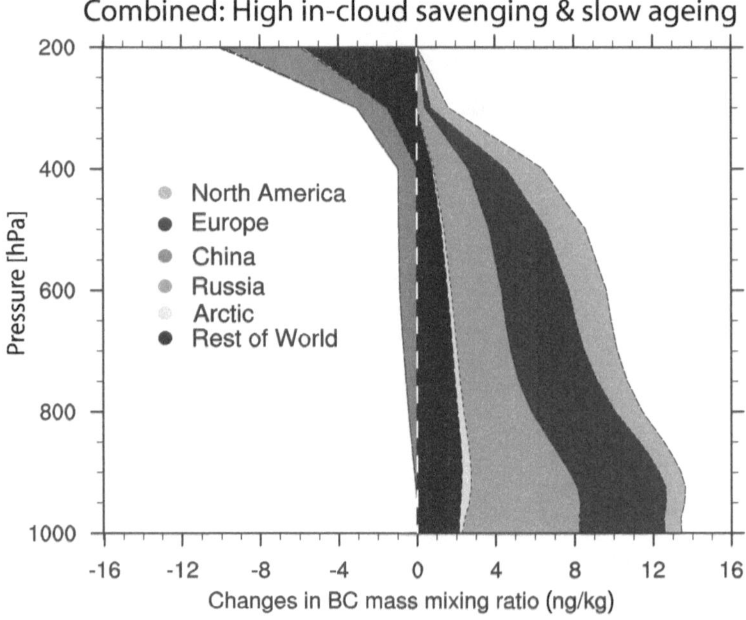

Figure 4.13.: As Figure 4.10, but for differences caused by changes in combinations of the in-cloud scavenging and ageing rate.

4.2.2. Two parameter sensitivity

Since the first part of the sensitivity study revealed that a combination of the parameters controlling the in-cloud scavenging and the particle ageing are capable of reducing the high-tropospheric high-bias in modelled BC concentrations, the effects are explored more in-depth here. This is done by varying the strengths of both parameters as discussed in subsection 2.3.2. For this, short 2-year runs without tagged tracers are sufficient. The factor scaling the in-cloud scavenging is denoted with α_{ic}. The one controlling the ageing by changing the required monolayer coating thickness is denoted with α_{ml}. For example, the run with combined parameters already discussed in section 4.2.1 is referred to as $Combined_{ic=2,ml=5}$ here.

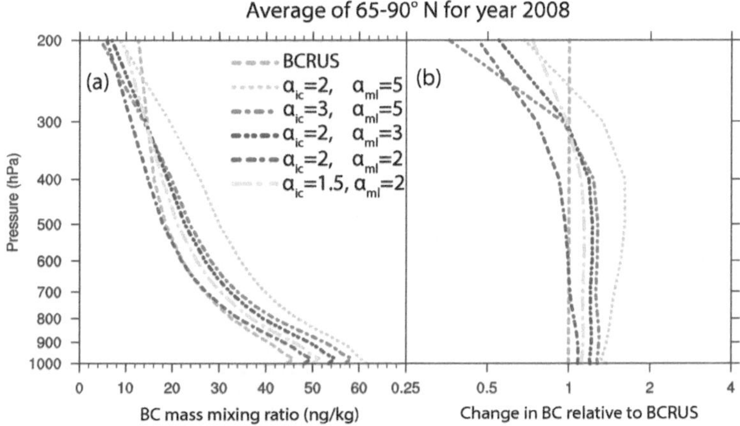

Figure 4.14.: Absolute values (a) and changes relative to the base run (b) in the annual mean vertical Arctic BC mass mixing ratios for the runs testing the sensitivity to the in-cloud scavenging efficiency (α_{ic}) and the required number of SU monolayers to age a particle (α_{ml}), for year 2008.

The setups were chosen in a way that the low and mid-tropospheric BC would be reduced when compared to $Combined_{ic=2,ml=5}$ while still lowering the high altitude BC compared to the base run $BCRUS$. This can be done by increasing the ageing rate compared to $Combined_{ic=2,ml=5}$, as in $Combined_{ic=2,ml=3}$, and $Combined_{ic=2,ml=2}$. The other option is to increase the in-cloud scavenging and therefore the wet deposition, as in $Combined_{ic=3,ml=5}$. In $Combined_{ic=1.5,ml=2}$ both are reduced.

The resulting annual average BC profiles north of 65° N can be seen in Figure 4.14a, with the changes relative to the base run $BCRUS$ shown in Figure 4.14b. The first thing to note is that the profiles change in the expected way, reducing the newly introduced low and mid-tropospheric high-bias, while addressing the high-tropospheric high-bias that was found in the evaluation. Interestingly, all of these runs again show the small maximum in the BC mass mixing ratio of roughly 950 hPa, that was visible for the base run. This is probably caused by the low Arctic boundary layer height, but was not visible for $Combined_{ic=2,ml=5}$, because the slow ageing increased the mid tropospheric BC mass mixing ratio too much.

$Combined_{ic=3,ml=5}$, shown in red in Figure 4.14, which deviates the strongest from the original model physics also shows the strongest effects at 200 hPa, with less than half the original amount of BC present compared to $BCRUS$. However, an increase of the in-cloud scavenging by a factor of three is a quite drastic change. This shows again that the model is more sensitive to an increase in this factor than to a change in the coating thickness required for ageing.

$Combined_{ic=2,ml=2}$, in purple, shows a similarly strong impact at 200 hPa, while also decreasing the BC mass mixing ratio above roughly 500 hPa. This run also exhibits the desired property of not deviating much in the mid-tropospheric BC compared to $BCRUS$. There is an additional upside in this version deviating the least from the original model physics. The slight increase in the required coating thickness is small compared to the value used in other models. Y. Wang et al. (2018) for example increase the coating parameter from 3 to 8 SU monolayers.

The run $Combined_{ic=1.5,ml=2}$, in grey in Figure 4.14, intends to change the model physics even less. It produces a BC mass mixing ratio profile, that is closest to $Combined_{ic=2,ml=2}$. The difference between the runs increases slightly with increasing altitude, again showing that the in-cloud scavenging is more important to the high tropospheric BC concentrations than to the near surface BC levels. The concentrations in $Combined_{ic=1.5,ml=2}$ are again higher than in the base run in the middle troposphere, therefore $Combined_{ic=2,ml=2}$ is chosen as the best estimate configuration and will later be referred to as $BC\text{-}Optimised$.

4.2.3. BC optimised setup

Arctic BC

The BC burden of this run optimised for the best agreement with Artic observed profiles is shown in Figure 4.15a. The differences between the BC burden of the $Base$ and this setup are shown in Figure 4.15b and the differences relative to the base run in c). It is very noticeable that the absolute and relative differences in BC burden are larger outside of the Arctic, which will be discussed in detail later. The absolute, as well as relative, differences inside of the Arctic are largest in the American part of the Arctic and around the Bering Strait, with more than $35\,\mu g\,m^{-2}$ less, which is a reduction by between 10 and 20 %. Over Greenland there is an decrease in the burden by more than 20 %, but only about $25\,\mu g\,m^{-2}$. These two regions are also where a statistically significant difference between the two burdens is found, following the criterion in section 2.5 which is stricter than a standard t-test at a 95 % confidence interval. Keeping in mind that the changes made to the microphysics affect the long range transport, while having a small effect on the more locally emitted BC, this result is not surprising. The European and Russian parts of the Arctic, while also affected by long range transport, are much more strongly affected by local sources.

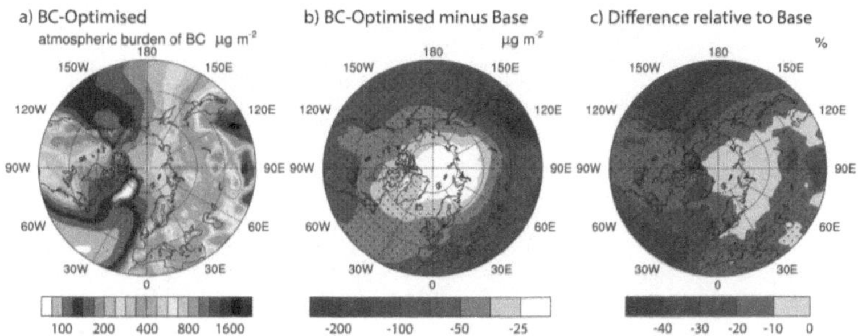

Figure 4.15.: (a) Arctic BC burden for *BC-Optimised* for the years 2008 to 2018. (b) Differences in the Arctic BC burden compared to *Base*. (c) Difference between *BC-Optimised* and *Base* relative to *Base*. The stippling shows regions of statistically significant differences (as defined in section 2.5) between the runs.

To evaluate the new setup designed to improve the Arctic BC mass mixing ratio profiles, it is compared against observations again, similar to section 3.2. Figure 4.16 shows the *Base* model setup in green, the *BC-Optimised* setup in purple and the observations in black. The thick lines again show the multi-campaign mean BC mass mixing ratios for each altitude calculated from all mean collocated measurement and model data points, for each season, respectively. The shaded area shows the minimum and maximum in this ensemble of campaign means. Still at least two different campaign instalments were used for each profile.

In autumn (see Figure 4.16a) one of the original problems of an overestimation between roughly 950 hPa and 550 hPa is still present. The biggest issue, which was the increase in modelled BC mass mixing ratios above 400 hPa, where observations showed a decrease, has mostly disappeared. Above 250 hPa, however, it still persists, even though at a lower intensity than in the base run.

The winter profile, shown in Figure 4.16b, has not changed drastically between *BC-Optimised* and the base run. This profile was well reproduced by ECHAM-HAM in the base setup already, and the problem of increasing mass mixing ratios with increasing altitude above roughly 350 hPa has disappeared. Like in autumn, there is still a slight overestimation of roughly $10 \, \mathrm{ng \, kg^{-1}}$ in the free troposphere.

As shown in Figure 4.16c, the *BC-Optimised* run produces slightly lower values in the area between 650 hPa and 350 hPa, where ECHAM-HAM already underestimated the average mass mixing ratios, compared to observations. The largest change is however to be found above that altitude. There, *BC-Optimised* produces BC mass mixing ratios that are closer to that of the observations than the base run.

For summer, the tendency for overestimation has slightly decreased, see Figure 4.16d. The expected and biggest improvements are to be found above 400 hPa, decreasing

Table 4.2.: BC lifetime in ECHAM-HAM.

Model run	Lifetime (d)	Wet dep. (kt yr^{-1})	Dry dep. (kt yr^{-1})
Base	6.86	8.35	0.88
BC-Optimised	5.55	8.40	0.85
BC-Optimised 3.4xGFAS	6.28	10.03	1.06

the overestimation from about a factor of three in the base run to about a factor of two in 250 hPa–200 hPa. The average profile for summer is still strongly affected by the ARCTAS-summer campaign where a missing precipitation event causes a strong overestimation in ECHAM-HAM (see section 3.2 for more details).

The modelled campaign profiles change as can be expected when looking at the relative changes for *BC-Optimised* shown in Figure 4.14. There is a strong decrease in high altitude BC, which allows for a more favourable comparison with the observations, combined with slightly reduced BC mass mixing ratios in the free troposphere. In the multi campaign averages this looks like a small setback, but the averages are especially in spring and summer influenced by strong biomass burning plumes that are dispersed in the model grid-box, instead of (possibly) more confined in reality, leading to a faster dispersion in the model. In winter *BC-Optimised* compares clearly more favourably. During this time there should be no biomass burning influence, and an even stronger decrease would be desirable.

In general, the changes in *BC-Optimised* make the already good agreement between ECHAM-HAM and airborne measurements of BC mass mixing ratios even better. The atmospheric lifetime changes from 6.86 d for *Base* to 5.55 d for *BC-Optimised* (see Table 4.2). This is still more than the often suggested lifetime of 5 d (Bauer et al. 2013; Q. Wang et al. 2014; Samset et al. 2014), but closer to it than before. Interestingly, the difference in the wet depositions of BC is pretty small between the two setups, with 8.40 kt yr^{-1} and 8.35 kt yr^{-1} for *Base* and *BC-Optimised*, respectively. It is obvious that the atmospheric lifetime of BC as the only reference point does not guarantee better results. The BC lifetimes for both setups are well within range of comparable models (see Table 1.1).

Other Arctic aerosol

Since these changes to the microphysics affect all aerosol in the model, and not only BC, as the focus of this work, the changes to other aerosol species will be examined here for the example of the ATom campaign measurements. This is done on the basis of all four instalments of ATom for OC and SU, and for three instalments for sea salt (SS), which was not availables for ATom-2. Figure 4.17 shows the Arctic profiles averaged between the instalments with the thick lines and maximum and minimum between the average instalment profiles as dashed lines.

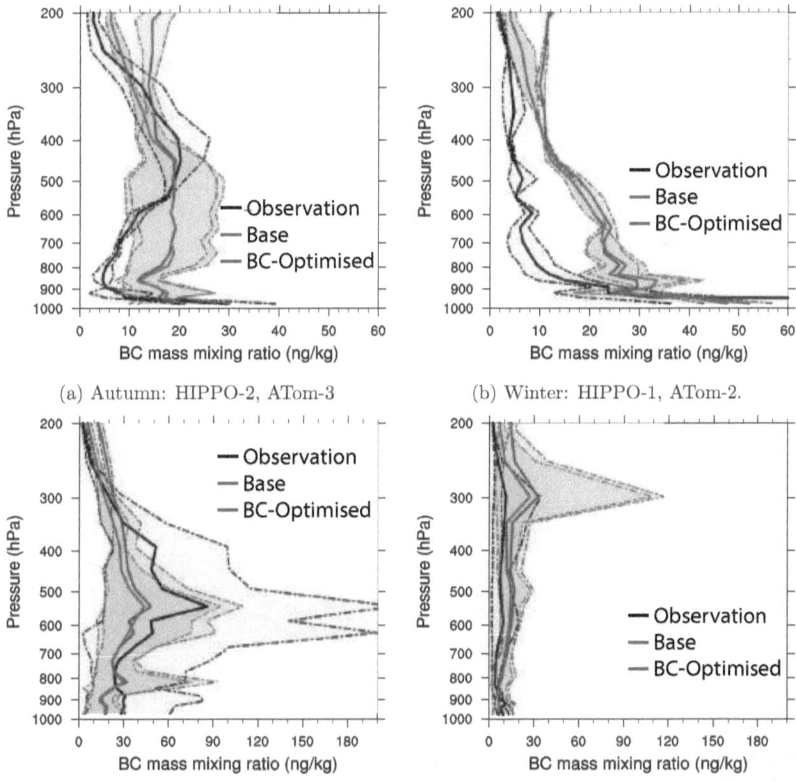

(a) Autumn: HIPPO-2, ATom-3

(b) Winter: HIPPO-1, ATom-2.

(c) Spring: ARCTAS-spring, ACLOUD, PA-MARCMiP, HIPPO-3, ATom-4.

(d) Summer: ARCTAS-summer, ACCESS, HIPPO-4, HIPPO-5, ATom-1.

Figure 4.16.: Vertical profile of the BC mass mixing ratios in each season as observed and modelled during the campaigns in that season. The average of the mean profiles from the campaigns is indicated by the bold line for observations and model in black and colour-coded, respectively. Green for run *BCRUS (Base)* and purple for BC-Optimised. The dashed line/shaded area show the maximum and minimum from the mean profiles.

The SU profile, shown in Figure 4.17a, shows a very good agreement for *Base* (in green) with the observations (in black) up to an altitude of 450 hPa. *Base* compares better to the observations than *BC-Optimised* (in purple), which underestimates the SU mass mixing ratio up to this altitude by about a factor of two. Above 450 hPa *Base*, however, overestimates the SU mass mixing ratio by about a factor of two, while *BC-Optimised* is closer to the observations. The behaviour of *BC-Optimised* generally producing lower SU mass mixing ratios is caused by the increase in in-cloud scavenging, leading to a faster wet deposition of SU.

The SS profiles shown in Figure 4.17b shows that both runs compare well to the ATom observations. The spread between the averages of the different instalments is, however, smaller in the model than observed. *Base* (in green) and *BC-Optimised* (in purple) show the biggest differences below 700 hPa. The SS mass mixing ratios are generally lower for *BC-Optimised*, which leads to a closer agreement between 900 hPa–700 hPa, while below it leads to a more favourable comparison between *Base* and the observations.

For OC the comparison is not as favourable as for the other species. As shown in Figure 4.17c, ECHAM-HAM underestimates the OC mass mixing ratio in both setups compared to the ATom observations. The observed profile shows a maximum in the averaged OC mass mixing ratios at roughly 530 hPa with 490 kg kg^{-1}. The averaged observed mass mixing ratio profile is relatively constant between 730 hPa–240 hPa, with lower values of around 300 ng kg^{-1} below and around 160 kg /kg above. *BC-Optimised* roughly captures this distribution, but shows an underestimation throughout the profile, with a maximum of 160 ng kg^{-1} in the averaged OC profile at 380 hPa. This underestimation is stronger than that of *Base*. The maximum of *Base* is found around 280 hPa with 280 ng kg^{-1}. It, however, shows an overestimation compared to the observations further aloft. This underestimation can at least partly be explain by ECHAM-HAM not considering volatile organic compounds (VOCs), which have the largest share of total OC emissions (Guenther et al. 2012; Iavorivska et al. 2016; Yang et al. 2019).

4.2.4. Global impact of optimising Arctic BC

While the focus of this work is the Arctic, ECHAM-HAM is a global model and it is most often used for global studies. It is therefore important to evaluate the impact of optimising the microphysics for Arctic BC profiles on the global aerosol population. In the latest global evaluation of ECHAM-HAM by Tegen et al. (2019) the model run *GFAS* compared the best against AERONET observations. It facilitated GFAS emissions, like most runs in this work, but with a global factor of 3.4 on the base values of the inventory, which was originally introduced to match global AERONET observations (Kaiser et al. 2012). In all other runs here this factor was discarded as it led to a strong overestimation in Arctic BC profiles, but is now used in the run *BC-Optimised 3.4xGFAS*, which is identical to *BC-Optimised* apart from this factor.

ECHAM-HAM is evaluated against AERONET observations in this section (as in Tegen

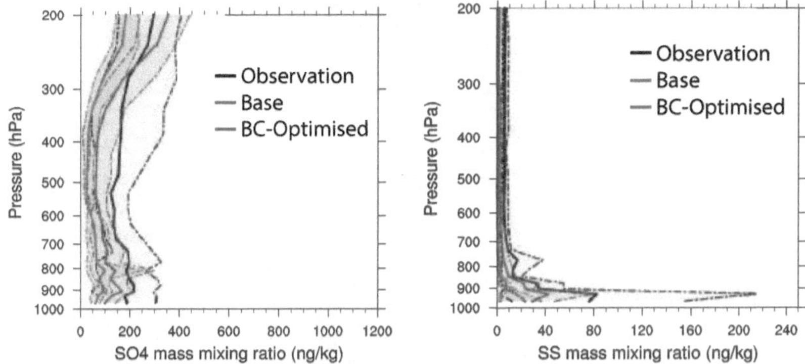

(a) SU mass mixing ratio: ATom-1, ATom-2, ATom-3, ATom-4

(b) SS mass mixing ratio: ATom-1, ATom-3, ATom-4

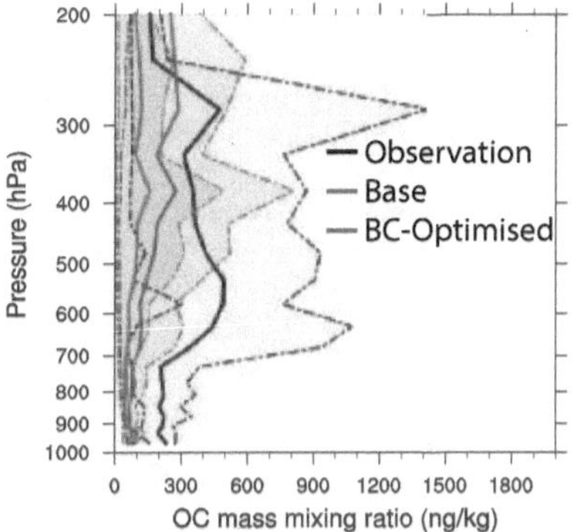

(c) OC mass mixing ratio: ATom-1, ATom-2, ATom-3, ATom-4.

Figure 4.17.: Vertical profile of the mass mixing ratios in the Arctic for different aerosol species as observed and modelled during the ATom campaigns. The average of the mean profiles from the campaigns is indicated by the bold line for observations and the model in black and green, respectively. The dashed line/shaded area show the maximum and minimum from the mean profiles.

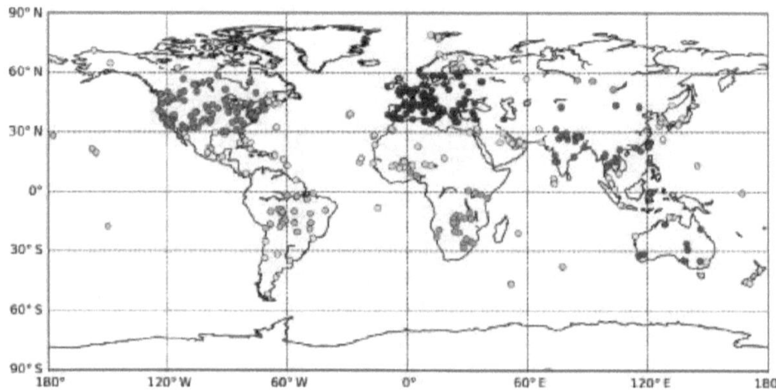

Figure 4.18.: AERONET stations colour-coded by region from Tegen et al. (2019). Red: North America; dark blue: Europe; brown: East Asia; pink: Siberia; yellow: North Africa; green: South Africa; orange: South America; dark green: Australia; light blue: oceanic regions; grey: elsewhere.

et al. 2019), as they give a standardised reference about the total aerosol with global coverage. In Figure 4.19 yearly mean AOTs of the AERONET network are plotted against collocated model values of the runs *Base*, *BC-Optimised*, and *BC-Optimised 3.4xGFAS*, respectively. The stations in Figure 4.19 are colour-coded by region as shown in Figure 4.18 (taken from Tegen et al. 2019). It shows a scatterplot of the modelled AOTs at 550 nm plotted against observations at 500 nm that were sampled from daily means which are collocated in space and time. The root mean square error (RMS) is similar for all three runs, but slightly larger for *BC-Optimised* with 0.13 compared to 0.12 for the other two runs. The RMS of all three runs here is in the range of the values found by Tegen et al. (2019). The normalised RMS, which is given in parentheses, is the lowest for *BC-Optimised* with 1.05, followed by *BC-Optimised 3.4xGFAS* with 1.09 and *Base* with 1.28, caused by the generally lower modelled values for the runs with changed microphysics. The AOTs of all three runs presented here are biased towards lower values. *BC-Optimised* and *BC-Optimised 3.4xGFAS* show a stronger negative (normalized) bias with -0.07 (-0.27) and -0.05 (-0.16), respectively, than *Base* with -0.04 (-0.04). The correlation, however, is better for those runs with 0.72 for *BC-Optimised* and 0.76 for *BC-Optimised 3.4xGFAS*. This behaviour of the highest correlation coefficient for the run that uses a 3.4 emission factor multiplied to the base emissions given in the GFAS emission data, is found by Tegen et al. (2019) as well. Tegen et al. (2019) also find that the model reproduced AOTs at sites with high AOT better when GFAS with the emission factor of 3.4 is used. Here, the effect of using GFAS with this factor in *BC-Optimised 3.4xGFAS* is smaller, especially at the remote stations (grey dots) with very low AOTs. This is very likely caused by the changes in the microphysics in *BC-Optimised* and *BC-Optimised 3.4xGFAS* that reduce the long range transport. The emission factor of 3.4 does, however, noticeably increase the modelled AOTs for the North American stations (red dots), bringing them closer to observed values.

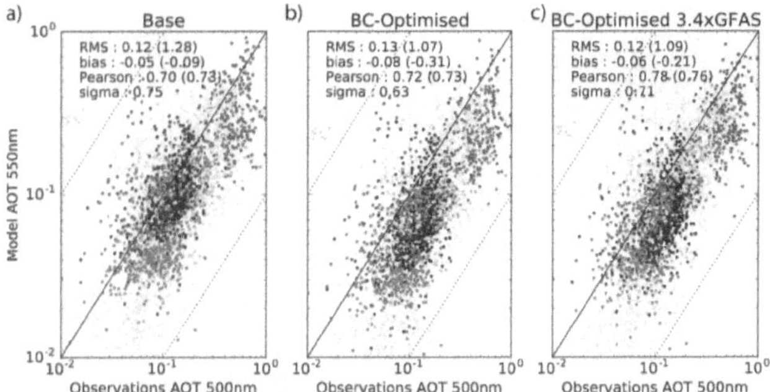

Figure 4.19.: Comparison of yearly means of modelled AOT values (at 550 nm) against observed AOTs (at 500 nm) at AERONET stations for the years 2007 to 2017. Data for the means was sampled from modelled daily means where a daily mean was available in the cloud screened level 2 AERONET product. The stations are colour-coded according to their regions (see Figure 4.18). For each comparison the root mean square error (RMS), the normalised RMS (in parentheses), the Pearson correlation coefficient on linear (Pearson) and on log scale (in parentheses), the absolute bias (bias) and normalised bias (in parentheses), and the ratio between the modelled and the observed standard deviation (sigma) are given.

Figure 4.20 shows a time series for monthly mean AOTs at a selection of stations (equivalent to Tegen et al. 2019). The observations are shown in black with grey bars that denote the standard deviation. For the different model runs only the mean is shown, in green, purple and brown, for *Base*, *BC-Optimised*, and *BC-Optimised 3.4xGFAS*, respectively. ECHAM-HAM is able to reproduce the seasonal cycles in all of the three runs reasonably well. The run with enhanced biomass burning emissions, *BC-Optimised 3.4xGFAS* is more capable to reproduce the peak concentrations in areas highly affected by biomass burning, as can be seen for Alta Floresta, Sao Paulo, and Lake Argyle. Both runs with the changed microphysics do not show the overestimation in spring that *Base* shows for the Goddard Space Flight Center (GSFC) (GSFC). For the other stations in Figure 4.20 the runs behave similarly.

A similar scatterplot as for the AOT is shown for the Ångstrom Exponent (AE) in Figure 4.21. The AE is a measure for the average effective particle size of aerosol particles in the atmospheric column. The AEs are derived from the measured extinction at 440 nm and 870 nm. For ECHAM-HAM they are derived from the AOTs at 550 nm and 865 nm. The model values and measurements are again collocated in time and space. The correlation between the model and measurements is lower than for the AOT with Pearson correlation coefficients between 0.49 and 0.52. The RMS is very similar between the three runs with 0.32 for *BC-Optimised* and *BC-Optimised 3.4xGFAS*, and

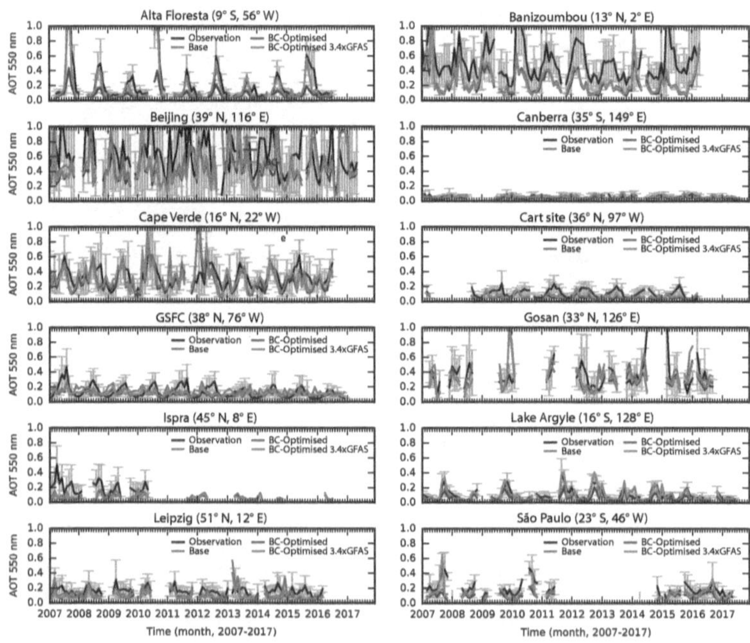

Figure 4.20.: Time series of monthly mean AOTs from January 2007 to December 2016 for a selection of AERONET stations (same set as Tegen et al. 2019). The black line shows the observations, with the error bars indicating the standard variability in daily mean measurements. The modelled values are sampled from daily means of the 12-hourly output, sampled on days where measurements are available. They are shown for the model runs *Base* (green), *BC-Optimised* (purple), and *BC-Optimised 3.4xGFAS* (brown).

0.31 for *Base*. The bias is the highest for *Base* with 0.07. A high bias in the AE means that the average size in aerosol particles is too high in this run. This means that the changes in the aerosol microphysics scheme for the other two runs is causing a higher proportion of small particles to be removed by wet scavenging. As also found by Tegen et al. (2019), the run with a 3.4 emission factor has a stronger positive bias than a comparable run without it. *BC-Optimised* shows the lowest bias at -0.01, compared to 0.04 that was found for *BC-Optimised 3.4xGFAS*. In the case here, this run is compared against a run using the GFAS emissions without this factor, instead of ACCMIP biomass burning emissions as in Tegen et al. (2019). Following their argument, this could mean that biomass burning emissions are too small in size in the model, which would lead to a high bias in the AE. However, it could also mean that the emission factor of 3.4 is causing this high bias, because it is leading to an overly high emission of (possibly correctly sized) small biomass burning particles.

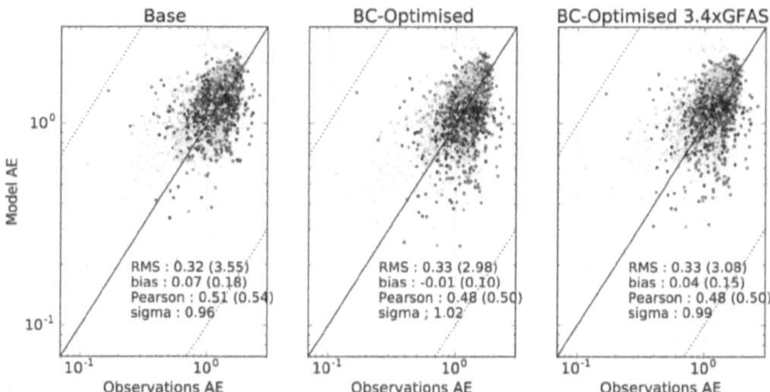

Figure 4.21.: Comparison of yearly means of modelled AE values with observed AEs at AERONET stations for the years 2007 to 2018. Data for the means was sampled from modelled daily means where a daily mean was available in the cloud screened level 2 AERONET product. The stations are colour-coded according to their regions (see Figure 4.18). For each comparison the RMS, the normalised RMS (in parentheses), the Pearson correlation coefficient on linear (Pearson) and on log scale (in parentheses), the absolute bias (bias) and normalised bias (in parentheses), and the ratio between the modelled and the observed standard deviation (sigma) are given.

The measurements from the ATom campaigns are especially suited for the global evaluation of an aerosol climate model like ECHAM-HAM, because measurements were performed without a sampling bias to meteorological or aerosol conditions. It covers the Pacific and Atlantic regions from 90°N to 70°S, allowing an evaluation of the capability of ECHAM-HAM to reproduce observed profiles of BC, OC, SU, and SS. Figure 4.22 shows one profile for each species, each covering the whole range from North to South Pole. The thick lines again show the average profile between the different campaign profiles available, with the shaded area showing the minimum between the single campaign averages and the maximum between the single campaign averages. Figure 4.22a shows that there is an improvement with *BC-Optimised* in the capability to reproduce the average BC profile above 600 hPa compared to *BC-Optimised 3.4xGFAS*. The BC mass mixing ratio improves from an overestimation by about a factor of five to an overestimation of three. This is mainly caused by a reduction in the northern mid-latitudes reducing the high bias there (not shown). Below the pressure level of 700 hPa both runs show a slight low bias of roughly 30 ng kg^{-1}, and 20 ng kg^{-1} for *Base* and *BC-Optimised*, respectively. *BC-Optimised 3.4xGFAS* agrees very well with the observations up to 750 hPa. Above that, however, there is a very strong high bias of almost one order of magnitude.

The average OC profiles for the ATom campaigns are shown in Figure 4.22b. The modelled multi campaign average of *BC-Optimised 3.4xGFAS* agrees well with the observational one, up to the pressure level of about 340 hPa, above which it exhibits an overestimation of a factor of two. The other two runs, *Base* and *BC-Optimised*, underestimate the OC concentrations. This low bias is even stronger for *BC-Optimised*. The difference in OC concentrations between *BC-Optimised 3.4xGFAS* and *BC-Optimised* explains in part the stronger low bias in the AOT of *BC-Optimised*.

The observed vertical SU profiles, black line in Figure 4.22c, cannot be reproduced well with any of the three setups. All of them show an overestimation by ECHAM-HAM, which was also found by Hodzic et al. (2019). Of the three runs, *Base* shows the strongest overestimation, which shows that the smaller low bias of *Base* in the AOT compared to that of *BC-Optimised 3.4xGFAS* could be an instance of errors canceling each other out.

For SS the average vertical profile is repoduced relatively well by ECHAM-HAM, as shown in Figure 4.22d. An overestimation can, however, be found for all runs below 700 hPa. The runs with the changed aerosol microphysics produce a smaller high bias below this level, with an overestimation of about a factor of two, compared to the factor of about three by *Base*.

4.2.5. Arctic AOTs and the 3.4 GFAS emission factor

The Arctic AERONET stations show an underestimation by ECHAM-HAM in the AOT values for all three runs, see Figure 4.23. The explanation for this is likely the underestimation in Arctic OC mass mixing ratios that can be seen in Figure 4.17c. The run with increased wildfire emissions from the GFAS emission set (*BC-Optimised 3.4xGFAS*, in brown) is better able to reproduce observed AOTs in the Russian Arctic, see Tiksi and Yakutsk. For some stations in the American Arctic, this run, however, produces large overestimations in case of high AOTs, see Bonanza Creek and Yellowknife Aurora. For these cases *Base* and *BC-Optimised*, that show a very similar behaviour in general, reproduce the observations more closely.

Figure 4.24 shows the average between all mean campaign profiles available in the Arctic. The run *BC-Optimised 3.4xGFA* is included in brown. This comparison, also the relation to the other runs, shows that by using the 3.4 emission factor on biomass burning emissions leads to a strong overestimation in the Arctic on average. This problem was not solved by the changes in the aerosol microphysics, as was the hope after making the changes. Thus, even though it was recommended by Kaiser et al. (2012) and leads to a better comparison with global AERONET measurements of the AOT, it is ultimately clear that it is not universally applicable. For work focussing on other regions than the Arctic and other aerosol species than BC, it might still be beneficial, even though it also lead to a worse comparison against the AERONET observations in terms of the AE. For these reasons, *BC-Optimised 3.4xGFAS* will not be considered further in this work.

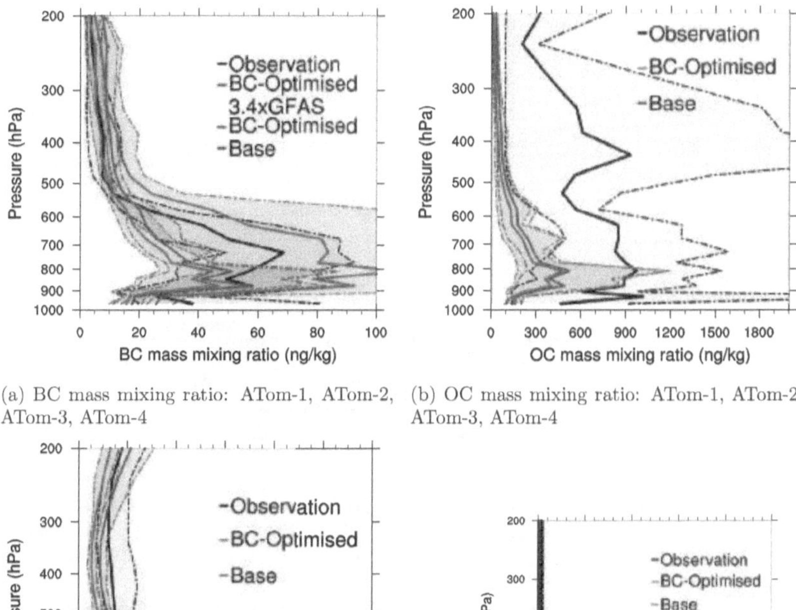

(a) BC mass mixing ratio: ATom-1, ATom-2, ATom-3, ATom-4

(b) OC mass mixing ratio: ATom-1, ATom-2, ATom-3, ATom-4

(c) SU mass mixing ratio: ATom-1, ATom-2, ATom-3, ATom-4.

(d) SS mass mixing ratio: ATom-1, ATom-3, ATom-4

Figure 4.22.: Vertical profiles of the mass mixing ratios in four different aerosol species as observed and modelled during the ATom campaigns. The average of the mean profiles from the campaigns is indicated by the bold lines for observations and for the model in black and green, respectively. The dashed line/shaded area show the maximum and minimum from the mean profiles.

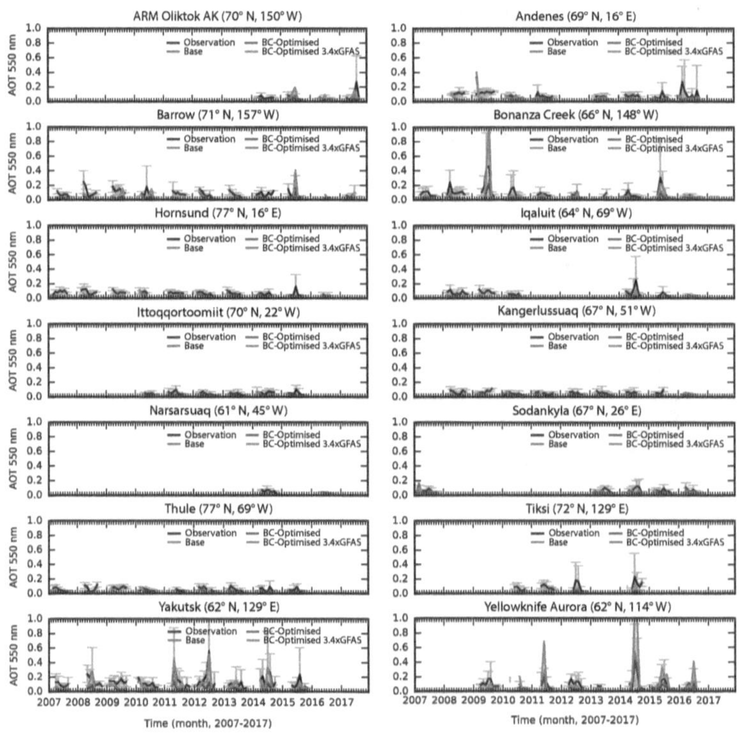

Figure 4.23.: As Figure 4.20, but for a selection of Arctic stations.

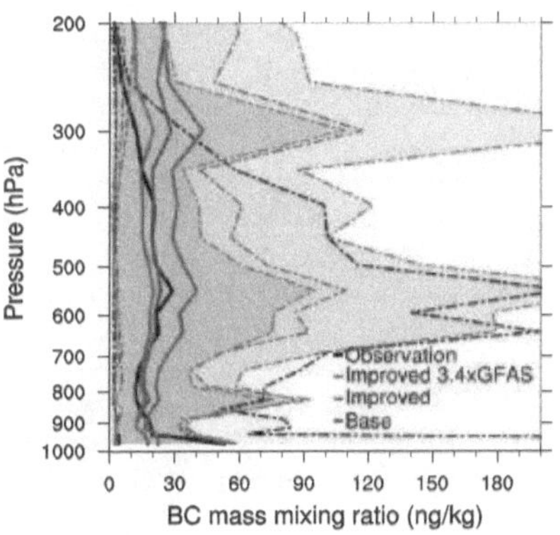

Figure 4.24.: As Figure 4.16, but for an average over all Arctic campaigns and including *BC-Optimised 3.4xGFAS* in brown.

Chapter 5.

Radiative effects of BC in the Arctic

To estimate the impact of BC on the Arctic climate system, the changes in the radiation fluxes at the top of the atmosphere (TOA) and the bottom of the atmosphere (BOA) are quantified. This is done for the DRE, semi-direct radiative effect (sDRE) and indirect radiative effect (IRE). The Arctic averages are calculated from values north of $60°$ N. The DRE of atmospheric BC is the main focus of this section, the BC-in-snow albedo effect which has just recently been implemented in ECHAM-HAM (Gilgen et al. 2018) is also considered as a contribution to this effect. The methods to calculate them are introduced in subsection 2.4.1 and subsection 2.4.2, for the DRE and the BC-in-snow albedo effect, respectively. For both of them an estimate is given for an unnudged setup first. The nudged runs are facilitated for another, constrained estimate and its uncertainties with respect to emissions and wet removal related microphysical processes are given. For the IRE and sDRE of atmospheric BC only an unnudged run estimate is given for one specific setup, as aerosol cloud interactions are not the main focus of this book. They are calculated as described in subsection 2.4.1. In addition to the multi-year annual means, the standard deviation among annual means is given in parenthesis as a measure of year to year variability.

5.1. Radiative estimate

The run *BC-Optimised* is used as the best estimate, as it proved to be best at reproducing vertical profiles of Arctic BC mass mixing ratios. However, the runs used in this section are unnudged, to not limit the atmospheric adjustments due to the BC radiative effects by constraining the run through nudging. As will be discussed in later sections, this noticeably affects the estimates of the DRE of atmospheric BC.

5.1.1. Clear sky DRE of BC

The multi-year mean average Arctic ($>60°$ N) DRE of BC for clear sky conditions at TOA is estimated at $0.20\,\mathrm{W\,m^{-2}}$ with an inter-annual variability (standard deviation of the annual means) of $0.14\,\mathrm{W\,m^{-2}}$. It is also shown in Figure 5.1a. For most of the Arctic

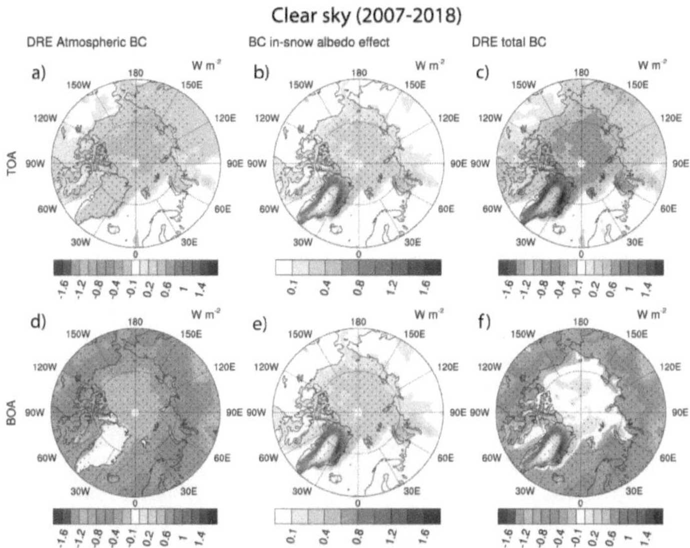

Figure 5.1.: Maps of multi-year (2007 - 2018) annual mean net radiative effects of BC in clear-sky conditions north of 60° N. From left to right the plots show the DRE of atmospheric BC, the BC-in-snow albedo effect, and the sum of the two. The top row and bottom row show TOA and BOA, respectively. The stippling shows statistical significance (as described in section 2.5)

Ocean it is higher than the Arctic average with more than $0.3\,\mathrm{W\,m^{-2}}$. This is caused by the absorbing effect of BC over the high albedo surface of sea-ice in this region, even though the BC burden in this region is lower than over the continental parts of the Arctic (see e.g. Figure 4.1). The DRE of atmospheric BC is still positive over most of the land area of the Arctic, but slightly negative (with around $-0.2\,\mathrm{W\,m^{-2}}$) over the Atlantic Ocean areas that largely stay ice free.

The stippling in Figure 5.1 shows the regions where BC statistically significantly alters the total monthly mean DRE of the total aerosol, at a level of $\alpha_{\mathrm{FDR}} = 0.05$, which is at least as strict as using a 95 % confidence interval, but becoming gradually more strict with each local significance test. It is calculated as described in detail in section 2.5 (Wilks 2016). A statistically significant difference is mostly given in the regions of the strongest DRE of BC, with some smaller regions in central Russia being the notable exceptions. The BC burden in those regions is often dominated by wildfire emissions, which change from year to year and have, therefore, no statistically significant impact on the multi year mean. However, they may contribute to a BC forcing trend over longer timescales.

Figure 5.1b shows the estimate of the radiative effect of BC-in-snow at TOA. This effect is only calculated for the solar range of the spectrum, to save computational resources, because the terrestrial effect of BC deposited on snow and ice is considered small. The multi-year annual Arctic mean (inter-annual variability) is estimated at $0.15\,\mathrm{W\,m^{-2}}$ ($0.01\,\mathrm{W\,m^{-2}}$). It is the strongest over glacier covered landmasses with the maximum of more than $0.9\,\mathrm{W\,m^{-2}}$ over Greenland. It is more than $0.2\,\mathrm{W\,m^{-2}}$ for most parts of the Arctic Ocean, that are covered by sea-ice most of the year. In this region, the BC-in-snow albedo effect also statistically significantly ($\alpha_{\mathrm{FDR}} = 0.05$) alters the TOA aerosol forcing, as shown by the stippling. For this, it is tested whether the total aerosol DRE changes statistically significantly when the BC-in-snow albedo effect is considered.

The maximum is again found over snow-covered glaciers, with more than $1.6\,\mathrm{W\,m^{-2}}$ over parts of Greenland. Over most of the Arctic Ocean the combined DRE and BC-in-snow albedo effect under clear sky conditions is between 0.4 and $0.8\,\mathrm{W\,m^{-2}}$. There the monthly means also statistically significantly alter the TOA total aerosol DRE.

At BOA the net DRE of atmospheric BC is negative with an Arctic average (inter-annual variability) of $-0.40\,\mathrm{W\,m^{-2}}$ ($0.13\,\mathrm{W\,m^{-2}}$), because less solar radiation reaches the surface. This shadowing effect causes a reduction in the net radiative flux of up to $-1.6\,\mathrm{W\,m^{-2}}$ at around 60°N over Russia. As shown by the stippling, BC statistically significantly alters the BOA DRE of the total aerosol almost everywhere, with Greenland as the most notable exception. Over the commonly sea-ice covered surfaces north of 75°N and over Baffin Bay, the reduction in net BOA irradiance is smaller than $-0.4\,\mathrm{W\,m^{-2}}$.

The impact of the BC-in-snow albedo effect is largely the same as for TOA (with an average and variability of $0.16\,\mathrm{W\,m^{-2}}$ and $0.01\,\mathrm{W\,m^{-2}}$, respectively) and has the same sign. Since the DRE of BC and the BC-in-snow albedo effect are of opposite sign at BOA, the influence of the combined effects is smaller, as shown in Figure 5.1f. There is a very clear division between a small net energy gain caused by BC over the often ice covered areas, like the central parts of the Arctic Ocean, Greenland and Nowaja Semlija, and the surrounding area with a net energy loss for the surface caused by BC.

5.1.2. All-sky DRE of BC

All-sky conditions, unlike the clear sky conditions, consider the presence of clouds, and so represent the actual atmospheric state of the Arctic. The DRE of BC in all-sky conditions is, therefore, much more relevant, since it reflects the impact on the climate. It is, however, harder to interpret and is of higher uncertainty since the relative position of clouds and BC layers are important.

The estimate for the DRE of atmospheric BC in all-sky conditions at TOA is $0.31\,\mathrm{W\,m^{-2}}$ on the multi-year annual mean. The inter-annual variability is $0.05\,\mathrm{W\,m^{-2}}$. This mean is also shown in Figure 5.2a. It is clearly shown that clouds amplify the net energy gain caused by atmospheric BC in the Arctic, because they increase the contrast of BC when they are located over dark water surfaces. The areas over the Atlantic that show a neutral impact or negative values for the clear-sky case show a net energy gain for all-sky conditions. The all-sky DRE of BC at TOA exceeds $0.4\,\mathrm{W\,m^{-2}}$ over most of the Arctic Ocean and the main parts of Siberia. It statistically significantly alters the DRE of all aerosol over most of the Arctic, which includes absorbing and reflecting aerosol.

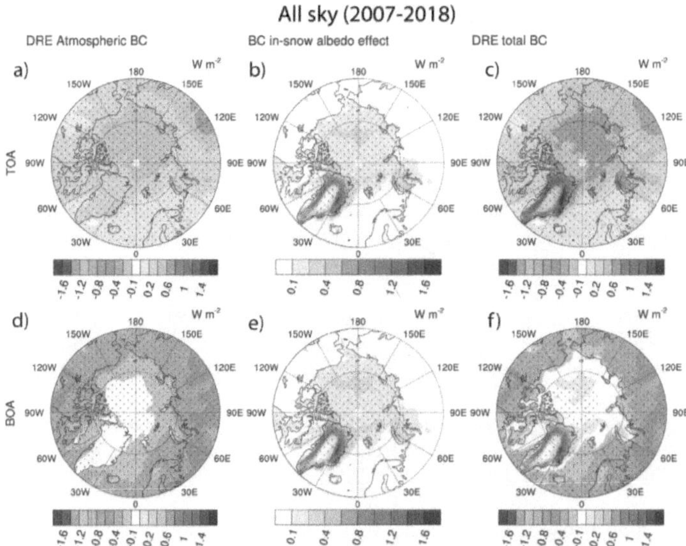

Figure 5.2.: As Figure 5.1, but for all-sky conditions.

At TOA, the BC-in-snow albedo effect, see Figure 5.2b, follows the same patterns for all-sky as for clear sky conditions. The effect is however about $0.1\,\mathrm{W\,m^{-2}}$ smaller, caused by less solar radiation reaching the surface because of the clouds.

Since the effect of atmospheric BC at TOA is more positive, while the effect of BC in snow is less positive in all-sky conditions than in clear-sky conditions, the sum of both is approximately equal over the central parts of the Arctic Ocean and the glacier covered land masses, as shown in Figure 5.2c. Over the more southern parts of the continents, however, the combined effect of the DRE of BC and the BC-in-snow albedo effect is stronger by about $0.1\,\mathrm{W\,m^{-2}}$ when considering clouds.

For the all-sky DRE it is also of interest whether it impacts the net radiative balance. This is, however, not the case. Even when adding the DRE and the BC-in-snow albedo effect, the impact is too small to find a statistical significance (not shown).

At BOA the DRE of atmospheric BC is slightly less negative than for clear sky conditions with a multi-year annual Arctic mean (inter-annual variability) of $-0.25\,\mathrm{W\,m^{-2}}$ ($0.6\,\mathrm{W\,m^{-2}}$), caused by the effect of shadowing clouds, which reduces the effect of BC. It is still the most negative over the continents at around $60°$ N (Figure 5.2d).

The BC-in-snow albedo effect is almost the same at BOA as at TOA, as is the case for the clear-sky case. On the Arctic average it is estimated at $0.12\,\mathrm{W\,m^{-2}}$ and $0.13\,\mathrm{W\,m^{-2}}$ at TOA and BOA, respectively. This effect still is the strongest over sea ice and glaciers, reaching up to $1.2\,\mathrm{W\,m^{-2}}$ over Greenland.

For the all-sky case at BOA there is a net energy gain over the central parts of the Arctic Ocean caused by the combination of atmospheric BC and the BC deposited in snow, as shown in Figure 5.2f. It is however below $0.2\,\mathrm{W\,m^{-2}}$. As the BOA DRE of BC is negative, the net energy gain is only caused by the BC-in-snow albedo effect. It is stronger over Greenland and other glaciated Arctic areas. In the rest of the Arctic, which consists of the continents, the Atlantic Ocean and the Pacific Ocean, it has a negative effect on the surface, which is only due to the shadowing effect of the atmospheric BC. These effects change the total aerosol DRE at BOA statistically significantly in most of the Arctic. On F^{net} there is no statistically significant impact.

As stated before, the Arctic averages are $0.31\,\mathrm{W\,m^{-2}}$ and $-0.25\,\mathrm{W\,m^{-2}}$ at TOA and BOA, respectively (see Table 5.1). At TOA this is slightly higher than the multi model median of the AeroCom models of 0.19, but within the model range (Sand et al. 2017). This is however not too surprising since they only report the change in DRE from pre-industrial levels. The value is a little lower than the possible reduction in total aerosol DRE for the Arctic found for a global reduction in BC emissions ($-0.4\,\mathrm{W\,m^{-2}}$, Kühn et al. 2020).

The additional or reduced energy that remains within the atmosphere by the respective effects are indicated by ATM in Table 5.1. It is calculated by subtracting the BOA from the TOA value. The DRE of BC, causes an additional $56\,\mathrm{W\,m^{-2}}$ to remain in the Arctic atmosphere on average. For the DRE this value is of interest since it represents is the amount of energy that is available to heat the air, and by that causes rapid adjustments.

On the global average the TOA DRE of BC is estimated at $0.08\,\mathrm{W\,m^{-2}}$, with an inter-annual variability of $0.09\,\mathrm{W\,m^{-2}}$. This is within the range of the AeroCom Phase II models. It is however notably lower than the multi model median of $0.23\,\mathrm{W\,m^{-2}}$ (Myhre et al. 2013b). This is even more notable because the AeroCom study only reported the DRE of the increase in BC from pre-industrial levels, as opposed to the DRE of total BC. However, if $BCRUS\text{-}NOBC$ is used as the reference run instead of $BCRUS\text{-}TRBC$, the corresponding DRE of BC is $0.38\,\mathrm{W\,m^{-2}}$. See subsection 2.4.1 for details on the technical differences. It is still significantly lower than the estimate at $0.71\,\mathrm{W\,m^{-2}}$ by Bond et al. (2013), which is at the high end of estimates.

The difference is even larger for the clear-sky case with $-0.11\,\mathrm{W\,m^{-2}}$ and $0.29\,\mathrm{W\,m^{-2}}$

Table 5.1.: All-sky multi-year (2007–2018) annual means (and inter-annual variability) in $\mathrm{W\,m^{-2}}$. Unnudged *BC-Optimised*. * IRE is only corrected for the pDRE (see subsection 2.4.1 for details) and the global TOA IRE should be stronger negative ($-0.13\,\mathrm{W\,m^{-2}}$, Cherian et al. 2017). The DRE of BC statistically significantly impacts the DRE of all aerosol. Neither of the three effects has a statistically significant impact on total net radiative balance of the Arctic.

	DRE		*IRE**		*sDRE*		Σ	
	Arctic	*Global*	*Arctic*	*Global*	*Arctic*	*Global*	*Arctic*	*Global*
TOA	0.31 (0.05)	0.08 (0.09)	-0.30 (1.30)	-0.01 (0.30)	-0.33 (1.19)	-0.08 (0.35)	-0.32	-0.01
BOA	-0.25 (0.06)	-0.97 (0.15)	-0.16 (0.60)	-0.01 (0.20)	0.09 (0.71)	0.19 (0.17)	-0.32	-0.79
ATM	0.56 (0.06)	1.06 (0.07)	-0.14 (1.30)	-0.01 (0.17)	-0.42 (1.19)	-0.28 (0.24)	0.00	0.77

for the methods using *BCRUS-TRBC* and *BCRUS-NOBC*, respectively. The difference is caused by an increase in the aerosol population of SU aerosol in *BCRUS-NOBC*. This increase shows up in the total aerosol DRE making the value more negative for *BCRUS-NOBC*. When calculating the difference between *BCRUS* and *BCRUS-NOBC*, this makes the difference positive. This means that additional reflecting SU in the *BCRUS-NOBC*, makes it seem as if there was more (or stronger absorbing) BC in the base run *BCRUS*. The main difference between the methods is located over the oceans and other dark surfaces, where especially aged BC has a negative TOA DRE. This is largely masked by the error introduced in the method where no BC is emitted. Over hotspot emission regions, where the freshly emitted BC largely is not coated, the DRE of BC is positive in any case. Plots of the global DRE of BC for both methods can be found in the appendix (Figure A.1, and Figure A.2, for both clear-sky and all-sky).

5.1.3. IRE and sDRE of BC

Both the IRE and sDRE are changes in cloud properties, caused by the presence of BC, which lead to changes in the radiative balance. They were computed with a further model experiment. The exact procedure is given in subsection 2.4.1.

Figure 5.3 shows the IRE and sDRE of BC. For both IRE and sDRE, TOA and BOA show similar patterns, unlike for the DRE where TOA and BOA were of opposite sign. The regional effects are stronger than for the DRE, showing a net energy gain in Greenland of more than $8\,\mathrm{W\,m^{-2}}$ and a net energy loss of more than $-8\,\mathrm{W\,m^{-2}}$ for the IRE and sDRE, respectively. These effects do not statistically significantly alter the net radiation (F^{net}) anywhere in the Arctic, using $\alpha_{\mathrm{FDR}} = 0.05$ for the criterion described in section 2.5, likely because of the stochastic nature of cloud occurrence. This variability in cloud occurrence affects the variance of the field compared against (F^{net}), meaning a statistically significant impact is hard to find. The DRE of BC that was compared against the DRE of total aerosol. These two significance tests should not be compared, because of the completely different reference fields.

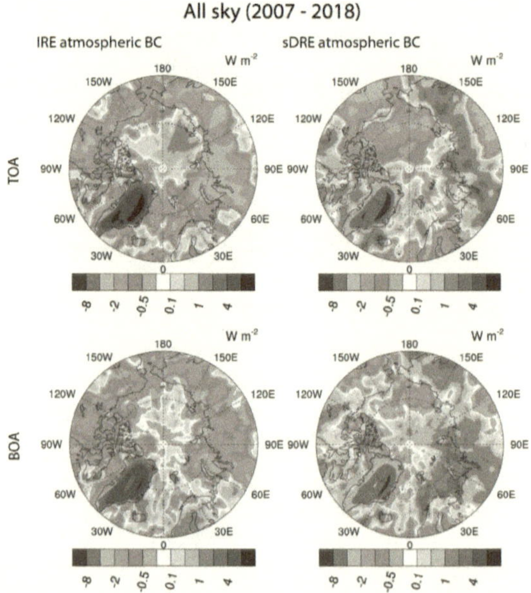

Figure 5.3.: Maps of annual mean net all-sky radiative effects of BC for the years 2007–2018. From left to right: IRE, and sDRE of atmospheric BC. Top row for TOA, bottom row BOA. No stippling in this figure indicates the lack of a statistically significant impact on the net radiative balance. Note the different levels for the colour map compared to the other radiative effect plots in this section.

IRE and sDRE are locally an order of magnitude higher than the DRE. At TOA they average out to $-0.30\,\mathrm{W\,m^{-2}}$ and $-0.33\,\mathrm{W\,m^{-2}}$ for the IRE and the sDRE, respectively, which then is on the order of magnitude of the DRE of BC. The inter-annual variability in the Arctic is an order higher than the means at $1.30\,\mathrm{W\,m^{-2}}$ and $1.19\,\mathrm{W\,m^{-2}}$ for the IRE and sDRE, respectively. At BOA the two effects additionally have an opposite sign, with Arctic averages (variabilities) of $-0.16\,\mathrm{W\,m^{-2}}$ $(0.60\,\mathrm{W\,m^{-2}})$ and $0.09\,\mathrm{W\,m^{-2}}$ $(0.71\,\mathrm{W\,m^{-2}})$ for the IRE and sDRE, respectively. In the Arctic the sDRE is caused a small reduction the total cloud cover. However, the vertically integrated cloud water increases slightly. The IRE is caused by a slight increase in both. The locatin and season of these changes in clouds relative to the underlying surface, as well as the seasona during which they occur would have to be analysed to understand these effects in detail.

The sum of the DRE, IRE, and sDRE of Arctic BC at TOA is $-0.32\,\mathrm{W\,m^{-2}}$, meaning that the cloud mediated effects more than offset the DRE. This behaviour is found by Kühn et al. (2020) for an experiment with BC emission reductions as well. They do, however, not give a value, since the variability is so high and changes in SU emissions as well as BC emission in their scenarios further complicate the clean separation of effects.

On the global average the TOA IRE of BC is estimated at $-0.01\,\mathrm{W\,m^{-2}}$ and the sDRE at $-0.08\,\mathrm{W\,m^{-2}}$. The estimate for the global TOA IRE of BC is, however, likely not strongly enough negative, because with the method used here, in one of the runs no BC is emitted, which causes the SU concentrations to change. The DRE of this increased SU is accounted for here, but not the effect it has on cloud properties (see also subsection 2.4.1). Cherian et al. (2017) use a more sophisticated method to arrive at a globally averaged IRE of BC of $-0.13\,\mathrm{W\,m^{-2}}$.

A study by Koch et al. (2010) gives an overview of different previous studies giving a range for the global "semi-direct" effect of BC at -0.4 to 0.1 $\mathrm{W\,m^{-2}}$, which is comparable to the combination of IRE and sDRE here at $-0.09\,\mathrm{W\,m^{-2}}$. With a global average of $-0.20\,\mathrm{W\,m^{-2}}$ Bond et al. (2013) estimate the combination of the effects at a similar strength. The combination of both was estimated to be higher and of opposite sign by Tegen et al. (2018) with a global average of $0.09\,\mathrm{W\,m^{-2}}$, also using ECHAM-HAM, but using a different emission and a different wet deposition scheme. These values, however strongly depend on the cloud regimes and vertical distributions of BC. An interpretation would therefore require an in-depth analysis.

5.2. Emission related uncertainty

To estimate the uncertainty in the DRE of BC with respect to anthropogenic emissions, the base run *BCRUS* is compared with the *ACCMIP-GFAS* run. This is done at TOA and BOA, but only for all-sky conditions, as they more accurately reflect the actual impact. Note that the runs are nudged unlike in section 5.1. This includes radiatively interactive and non-interactive BC model runs. A statistically significant difference between the two runs is easier to detect because of the common meteorological constraints.

Again, the base run *BCRUS* is characterised by updated emissions, consisting of ECLIPSE (Klimont et al. 2017) with increased Russian BC (K. Huang et al. 2015), while *ACCMIP-GFAS* uses fixed year 2000 anthropogenic emissions instead. The BC emissions of *ACCMIP-GFAS* are, therefore, lower in north and central Asia, but higher in North America and Europe, ultimately leading to a lower Arctic BC burden for *ACCMIP-GFAS* (see Figure 4.1). Both use satellite based daily changing biomass burning emissions from GFAS (Kaiser et al. 2012).

5.2.1. TOA

As already discussed in section 5.1, the multi-year net annual mean DRE and the BC-in-snow albedo effect cause a net energy gain for the Arctic system. For the base run the Arctic average of the DRE of atmospheric BC is $0.37\,\mathrm{W\,m^{-2}}$, with the in-snow-BC albedo effect at $0.13\,\mathrm{W\,m^{-2}}$. The DRE of atmospheric BC is more relevant over most of the area, but the BC-in-snow albedo effect is stronger over glaciers, especially on the slopes of Greenland, as shown in Figure 5.4a-c. The stippling shows a statistically significant difference caused by the respective effect, according to the criterion described in detail in section 2.5 (Wilks 2016).

Figures 5.4d-f show the differences between *BCRUS* and *ACCMIP-GFAS*. The differences in the multi-year annual mean DRE of atmospheric BC caused by using these fixed emissions is less than $0.1\,\mathrm{W\,m^{-2}}$ in most areas of the Arctic. Only in some regions of the Russian Arctic, that already showed the biggest changes in the atmospheric burden of BC (compare Figure 4.1), show slightly larger differences. It is, however, notable that the slightly larger differences occur over the Kara Sea and Laptev Sea, which are considered to be particularly sensitive for the connection between Arctic sea-ice loss and changes in the large-scale atmospheric circulation (e.g., Petoukhov et al. 2010)). Very locally in a gas flaring region, values of up to $0.25\,\mathrm{W\,m^{-2}}$ can be found. This means that *BCRUS* has a more positive DRE in the Arctic, especially in the gas flaring regions.

The Arctic average in DRE of atmospheric BC for *ACCMIP-GAS* is $0.29\,\mathrm{W\,m^{-2}}$, which is $0.08\,\mathrm{W\,m^{-2}}$ less than for *BCRUS*, leading to an uncertainty estimate of about 25%. The difference in DRE of atmospheric BC due to anthropogenic emission uncertainties is, however, not statistically significant in the Arctic. For the combination of the effects the net radiation considering both is compared to the net radiation considering neither, to test the significance. For differences between the model runs on the bottom row, the monthly mean fields for the respective variables are compared.

The differences in the BC-in-snow albedo effect at TOA are equally small (see Figure 5.4e). There are areas in Greenland where the effect increases, likely caused by the higher amounts of European BC in *ACCMIP-GFAS* deposited on the slopes of Greenland. The largest reduction is found in the west of the Kara Sea, where in the *BCRUS* run considerably more BC occurs in the gas flaring region, adding to the effect of the atmospheric BC. The Arctic average reveals an uncertainty of 15% for the BC-in-snow albedo effect at TOA.

The uncertainty introduced by different emission configurations is about 25% at TOA when considering both the DRE and the BC-in-snow albedo effect, but is not statistically significant.

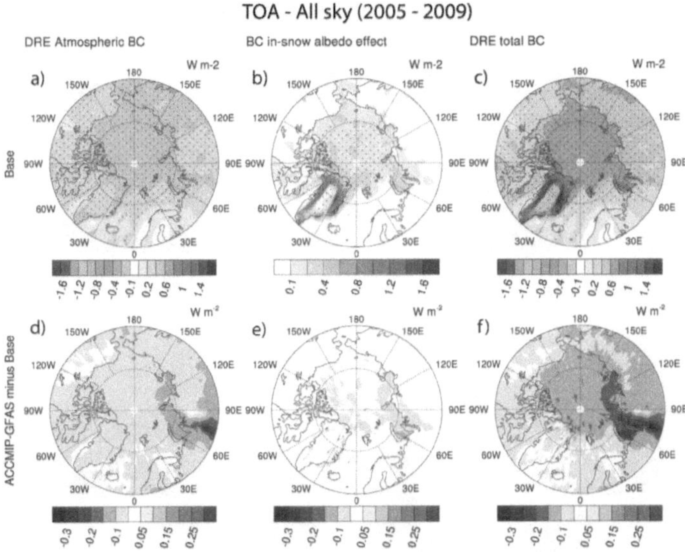

Figure 5.4.: TOA DRE, and in-snow albedo effect, as well as respective uncertainty caused by uncertainty in emissions. Maps of annual mean net all-sky radiative effects of BC for the years 2005-2009. The stippling shows statistical significance.

5.2.2. BOA

The DRE of atmospheric BC is negative at BOA, as discussed in subsection 5.1.2. Figure 5.5a shows this for the nudged $BCRUS$ run, with around $-0.1\,\mathrm{W\,m^{-2}}$ over the Arctic Ocean, increasing in strength southward to around $-0.8\,\mathrm{W\,m^{-2}}$ over Europe and Asia. The regions where BC has a statistically significant impact on the BOA DRE of all aerosol types is indicated by the stippling. The BC-in-snow albedo effect is causing a net energy gain at BOA, as also discussed earlier, and statistically significantly alters the BOA DRE of all aerosol.

The combined effects show net energy gain for glaciated regions e.g. Greenland and the sea-ice north of Svalbard, there they alter the DRE of all aerosol statistically significantly. There is net energy loss by the combined effects over the continents and the Atlantic Ocean and Pacific Ocean, with roughly $0.2\,\mathrm{W\,m^{-2}}$ and $-0.4\,\mathrm{W\,m^{-2}}$, respectively, which also affect the DRE of all aerosol statistically significantly. Also for $BCRUS$ the most positive effect at BOA is found in Greenland with an energy gain of more than $1.4\,\mathrm{W\,m^{-2}}$ locally.

In contrast to the base run $BCRUS$, the run with fixed year 2000 anthropogenic emis-

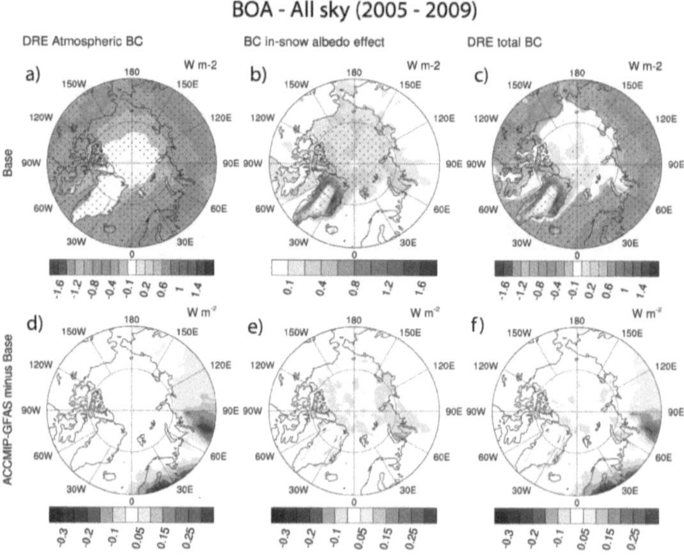

Figure 5.5.: As in Figure 5.4, but for BOA.

sions, *ACCMIP-GFAS*, produces an even larger negative DRE of atmospheric BC in northern Europe, where more BC is emitted in this run, as shown in Figure 5.5d. The negative DRE is lower by more than $0.3\,\mathrm{W\,m^{-2}}$ locally in the gas flaring regions in Russia. In the centre of that region a statistically significant difference between the DRE of BC in *BCRUS* and *ACCMIP-GFAS* is detected, with the false discovery rate (FDR) limiting method described in section 2.5.

The emission related uncertainty by the BC-in-snow albedo effect at BOA, shown in Figure 5.5e, is roughly the same as for TOA. Unlike for BOA it is however more important north of 75°N than the DRE of atmospheric BC.

5.3. Wet removal related uncertainty

Two setups are used to estimate the uncertainty in the DRE of BC and the BC-in-snow albedo effect caused by an uncertainty in the wet removal. Both facilitate the *BCRUS* emission setup. The run called *BCRUS* previously is, therefore, now referred to as *Base*. *BC-Optimised*, which combines an adjustment of the in-cloud scavenging parametrisation and a slower ageing of aerosol particles, is compared against *Base*. The DRE of BC for each is calculated by using two different runs, with the only

Table 5.2.: Multi-year (2005–2009) annual means (inter-annual variability) in $\mathrm{W\,m^{-2}}$. Nudged *BCRUS* and emission related uncertainty (Δ, *BCRUS* minus *ACCMIP-GFAS*). No statistically significant differences between the monthly Arctic means were found at the 95 % confidence level.

	DRE		in-snow		Σ	
	Base	Δ	Base	Δ	Base	Δ
TOA	0.37 (0.08)	0.08	0.13 (0.02)	0.02	0.50	0.10
BOA	-0.32 (0.02)	-0.02	0.14 (0.02)	0.02	-0.18	-0.00
ATM	0.68 (0.10)	0.10	-0.01 (0.00)	-0.00	0.68	0.10

difference between them being that BC does not interact with radiation in one of them (as described in detail in subsection 2.4.1). The runs used in this section are nudged, unlike the runs in section 5.1.

5.3.1. TOA

The top row of Figure 5.6 shows the TOA DRE of BC and BC-in-snow albedo effect for *BC-Optimised*. Both effects are positive (warming) over the Arctic, with a DRE of BC of more than $0.4\,\mathrm{W\,m^{-2}}$ for most of the Arctic Ocean, and a BC-in-snow albedo effect of more than $0.1\,\mathrm{W\,m^{-2}}$. Together they amount to more than $0.6\,\mathrm{W\,m^{-2}}$ for most of the central Arctic. The maximum of the sum of both is found over the coastal areas of Greenland with more than $1.6\,\mathrm{W\,m^{-2}}$, because of the high BC-in-snow albedo effect in these areas. In the multi-year annual Arctic average the TOA DRE of BC amounts to $0.31\,\mathrm{W\,m^{-2}}$ for *BC-Optimised*.

The bottom row of Figure 5.6 shows the difference with respect to *Base*. While the BC-in-snow albedo effect shows to be largely the same, the DRE of BC in *BC-Optimised* is somewhere between $0.1\,\mathrm{W\,m^{-2}}$ and $0.2\,\mathrm{W\,m^{-2}}$ lower than in *Base*. The difference in the DRE of BC is also statistically significant over the sea-ice regions of the American Arctic, as well as Greenland. It is also statistically significant over the Atlantic, where the difference between the runs is very small.

Even though this change in DRE is related to a difference in the BC burden, the burden is only an indicator. The central Arctic Ocean, as a region with the highest absolute changes in TOA DRE of BC, was the region with the lowest absolute change in BC burden, as shown in Figure 4.15. Even more interesting is the fact that the larger difference in BC burden over the central Arctic Ocean between *ACCMIP-GFAS* and *Base* causes a smaller shift in the TOA DRE. This shows the importance of the position of BC relative to clouds and to a lesser extent other aerosol. A much bigger part of the reduction in BC burden caused by the change in aerosol microphysics comes from a reduction in high altitude BC, unlike the change caused by a difference in the emissions, which caused a much more uniform change in concentrations throughout the atmospheric column (see subsection 4.1.5 and section 4.2.3).

The Arctic average of the DRE of atmospheric BC for *BC-Optimised* is $0.10\,\mathrm{W\,m^{-2}}$ lower than in *Base*. The uncertainty estimate caused by uncertainties in wet deposition of BC is, therefore, very substantial at about 25 %. This uncertainty caused by an uncertainty in wet removal is therefore in the range of that caused by emissions. The affected region is however different. The uncertainty from emissions affects the Eurasian Arctic coastal regions more strongly, while the uncertainty from the wet deposition affects the high Arctic around the North Pole more strongly.

Table 5.3.: Multi-year (2007–2018) annual means (inter-annual variability) in $\mathrm{W\,m^{-2}}$. Nudged *BC-Optimised* and wet removal related uncertainty (Δ, *BC-Optimised* minus *Base*). **Bold** values show statistically significant differences between the monthly Arctic means at 95 % confidence.

	DRE		*in-snow*		Σ	
	BC-Optimised	Δ	*BC-Optimised*	Δ	*BC-Optimised*	Δ
TOA	0.31 (0.04)	**-0.10**	0.13 (0.01)	0.00	0.43	-0.10
BOA	-0.24 (0.03)	**0.11**	0.13 (0.02)	-0.00	-0.11	0.10
ATM	0.55 (0.07)	**-0.21**	0.00 (0.00)	-0.00	0.55	-0.21

BOA

The DRE of atmospheric BC on the BOA for the unnudged *BC-Optimised* run is shown in Figure 5.7a. It causes a net energy loss for the Arctic surface that is less than $-0.1\,\mathrm{W\,m^{-2}}$ around the North Pole, but becomes increasingly important on the continents with values more negative than $-0.3\,\mathrm{W\,m^{-2}}$. The BC-in-snow albedo effect (Figure 5.7b) is of opposite sign, causing a net energy gain on the surface. The combination of both effects leads to a net energy gain caused by the presence of BC for the commonly ice-covered surfaces of the high Arctic and Greenland, while causing a net energy loss at BOA south of this.

The difference between the runs *Base* and *BC-Optimised* that represents the uncertainty related to aerosol microphysics, is shown in the lower part of the panel in Figure 5.7. Since the BC-in-snow albedo effect barely changes at BOA as well as was the case for TOA, it is dominated by the difference in DRE by atmospheric BC (shown in d). The multi-year Arctic annual average BOA DRE of atmospheric BC of *BC-Optimised* shows a substantially weaker net energy loss for BOA, at only $-0.24\,\mathrm{W\,m^{-2}}$, compared to *Base* at $-0.35\,\mathrm{W\,m^{-2}}$. This leads to an uncertainty estimate at BOA of roughly 30 %, which is slightly higher than at TOA. The uncertainty is less than $0.05\,\mathrm{W\,m^{-2}}$ around the North Pole, caused by the generally low impact on the net radiative flux at BOA caused by atmospheric BC in this region. Over the continental Arctic the

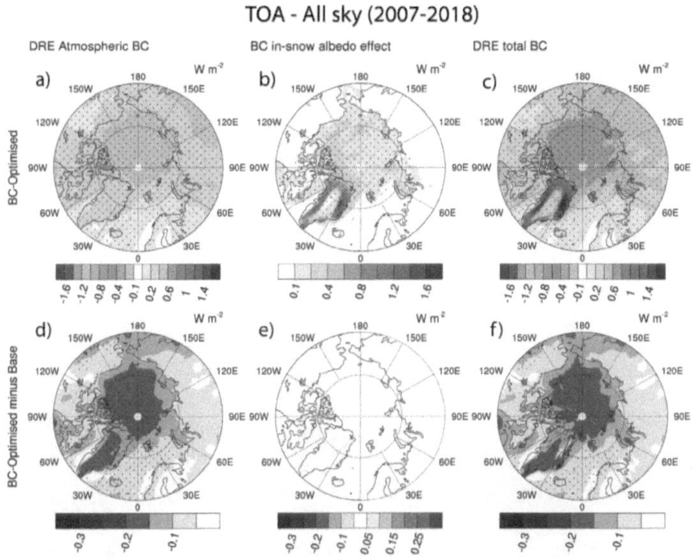

Figure 5.6.: Uncertainty in TOA DRE and in-snow albedo effect caused by an uncertainty in the wet deposition of BC. Maps of annual mean net all-sky radiative effects of BC for the years 2007–2018. The stippling shows statistical significance.

uncertainty is, however, more than $0.10\,\mathrm{W\,m^{-2}}$. This difference between the setups is largely statistically significant according to the FDR limiting criterion described in section 2.5.

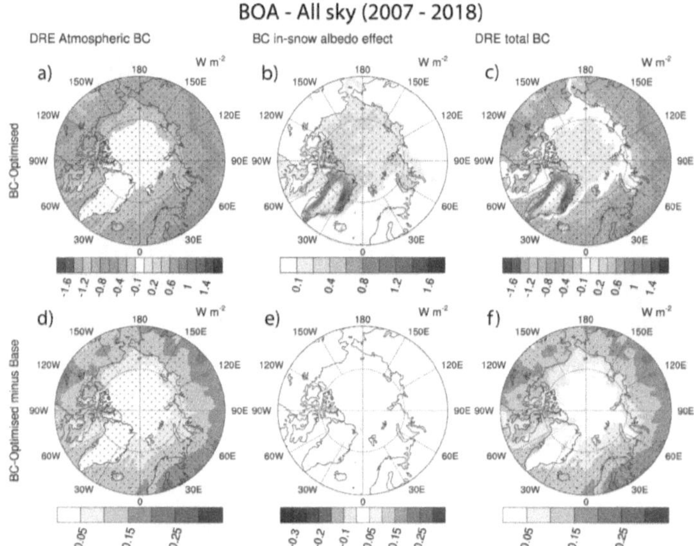

Figure 5.7.: As in Figure 5.4, but for BOA.

Chapter 6.

Summary and Outlook

This book covers the modelling of Arctic black carbon (BC) and explores the radiative impact of BC, which is known to be highly absorbing in the visible light spectrum. BC is, therefore, a potentially important contributing factor in the rapidly warming Arctic climate. The known key uncertainties in modelling Arctic BC were focussed on, which are the seasonality of occurrence and the vertical distribution of BC. The main sources for these key uncertainties were analysed. These are the emissions and the ageing and wet removal during long-range transport of BC.

The main aim was to give a state-of-the-art estimate of the distribution and radiative effects of Arctic BC, which is based on recent emission data and model developments, and is supported by the most recent Arctic observational efforts. This entails evaluating how strongly the global aerosol climate model ECHAM-HAM is affected by the uncertainties in ageing and removal during long-range transport, as well as the uncertainties in emissions. This work therefore provides the first quantification of the model uncertainty for the distribution and, in particular, the direct radiative effect (DRE) of BC aerosol in the Arctic related to these sources of uncertainty.

To reach these goals, ECHAM-HAM was evaluated against a comprehensive set of observational data from airborne campaigns as well as surface sites. This was followed by sensitivity studies on the main sources of model uncertainty with regard to the sources and sinks of Arctic BC and relevant model parameters controlling the ageing and wet deposition during long-range transport.

First, the emission-related uncertainties were tested with four different setups, consisting of state-of-the-art emission datasets. The setups were chosen, so that the differences between the model runs show the following uncertainties or benefits: (1) Using daily changing satellite-based based (GFAS), instead of climatological biomass burning emissions; (2) fixed anthropogenic emissions of the year 2000 (ACCMIP) versus up-to-date, time-evolving emissions (ECLIPSE v5a); (3) regional, spatially higher resolved update of Russian emissions with a strong contribution of gas flaring, where the anthropogenic ECLIPSE BC emissions were replaced by the regional emission dataset of K. Huang et al. (2015). As a result, a new emission configuration was created, suited to study the distribution and radiative effects of BC in the Arctic region. This setup, *BCRUS*, consists of GFAS biomass burning emissions and ECLIPSE v5a anthropogenic emissions, and refined BC emissions over Russia, considering important gas-flaring sources.

Secondly, the sensitivity to different factors controlling the wet deposition of BC was tested in order to give an estimate of the long-range transport related uncertainty. In the first step, the relevant processes were investigated individually: (1) The speed of the aerosol ageing was altered by changing the required number of sulphate (SU) monolayers, (2) the in-cloud scavenging efficiency was scaled up and down, and (3) the SU emissions were varied, controlling the availability of the SU needed for the ageing of aerosol particles. Finally, combinations of slower ageing and increased in-cloud scavenging were iterated to find a setup, "*BC-Optimised*", which better reproduces observed Arctic BC profiles. For this the model was upgraded with BC tracers that are tagged by their respective source region. There exist only few tagging studies for Arctic aerosol. Here for the first time tagging was used for optimisation of the parametrisations related to wet removal, allowing a more in-depth analysis on how different transport pathways are affected.

The initial evaluation of ECHAM-HAM revealed that the model is largely capable of reproducing local seasonal cycles of the BC mass concentration in the Arctic. The vertical distribution, however, proved to be more challenging, revealing a clear tendency to overestimate BC mass mixing ratios in the upper troposphere. In the levels above roughly $300\,\mathrm{hPa}$, a decrease in BC with increasing altitude can be typically observed, while modelled concentrations showed a tendency to increase with altitude in this study. Below this level, however, ECHAM-HAM performed well. So far, hardly any other study has so thoroughly evaluated the vertical distribution of BC in the Arctic throughout the year and used these results to systematically optimise the model with sensitivity studies.

In the first sensitivity study on BC emissions, it was shown that satellite-based, daily changing biomass burning emissions are crucial to reproduce the observed aerosol vertical distribution and considerably increase the correlation between modelled and observed values at Arctic research sites. While the differences between the setups can be large, the ones using GFAS do perform better or worse depending on the meteorological conditions and geographical location, with no clear advantage for any one of them. The differences between all setups can be large, but for those that were using GFAS all setups show cases where they perform best, depending on the geographical location and meteorological conditions. On the monthly average, even the run with monthly changing biomass burning emissions that were fixed for a year performed reasonably well. The largest uncertainty in the Arctic BC burden of over $100\,\mathrm{\mu g\,m^{-2}}$ (about $25\,\%$) was found in the Russian Arctic, where observations are the scarcest. The emission setup *BCRUS* was then chosen for further research, on the merits of performing the best in comparison to the monthly means of observations at the only Russian Arctic station available.

In the second sensitivity study on factors relevant for the wet removal of BC, the sensitivity of ECHAM-HAM proved to be high to parameters controlling the ageing of aerosol particles and in-cloud scavenging. The sensitivity of the BC concentrations to an enhanced in-cloud scavenging proved to be strongly altitude-dependent, with by far the largest reduction in upper-tropospheric BC. The sensitivity to a slower ageing, however, showed more of a uniform increase in concentrations throughout the vertical profile. These two parameters were, therefore, combined in a setup with a twice as efficient in-cloud scavenging and a slower ageing requires double the amount of SU to age an aerosol particle. This setup proved to efficiently reduce the high bias in high-altitude BC. Moreover, it improved the vertical profiles of BC globally, and for SU in the lower troposphere, when compared against ATom aircraft measurements. The global lifetime of BC decreased from 6.9 d to 5.6 d. The biggest impact on the Arctic BC burden was located over Asia and North America around the Bering Strait, but the reduction compared to the base run was smaller than $50\,\mu\mathrm{g\,m}^{-2}$ throughout the Arctic. This impact on the average Arctic BC burden was smaller than the one found for the sensitivity study on BC emissions, with an uncertainty of $26\,\mu\mathrm{g\,m}^{-2}$ ($10\,\%$) compared to $63\,\mu\mathrm{g\,m}^{-2}$ ($25\,\%$).

From the synthesis of the two sensitivity studies followed, that Russian gas flaring and increased BC emissions resulting from economic growth mainly in East Asia since the year 2000 have lead to an increase in the Arctic BC burden by $15\,\%$, even though emission have decreased in Europe, North America and Japan. The tagging of BC tracers revealed that Chinese and other long-range transported BC contributed especially to upper tropospheric concentrations. It can thus be concluded, that the upper-level long-range transport has become more important. However, it was found that the impact of this long-range transport on the Arctic, while still relevant, was originally overestimated in the base version of ECHAM-HAM, most likely because of a too low in-cloud scavenging efficiency.

To give estimates for the radiative effects of BC in the Arctic, the radiative transfer code in ECHAM-HAM was upgraded with the option to consider BC (or any other aerosol species) as transparent in the calculation of the aerosol-radiation interactions. This means that BC was still available to act as cloud condensation nuclei or as ice nucleating particles. It therefore enabled the calculation of the DRE, the semi-direct radiative effect, and the indirect radiative effect (IRE) of BC from a set of model runs. Using a recent upgrade to the radiative transfer code of ECHAM-HAM also enabled the consideration of the BC-in-snow albedo effect (Engels 2016; Gilgen et al. 2018).

Globally, the total annual top of the atmosphere (TOA) effective radiative effect of atmospheric BC was estimated at $-0.01\,\mathrm{W\,m}^{-2}$ in this study. This value comprises of a DRE of $0.08\,\mathrm{W\,m}^{-2}$, an IRE of $-0.01\,\mathrm{W\,m}^{-2}$ and a semi-direct radiative effect (sDRE) of $-0.08\,\mathrm{W\,m}^{-2}$. The estimate for the DRE of BC is considerably lower than the multi-model median from the AeroCom model intercomparison initiative of $0.23\,\mathrm{W\,m}^{-2}$, but within the multi-model range (Myhre et al. 2013b). The combination of the IRE and sDRE of BC is very close to what is found in the literature with $-0.09\,\mathrm{W\,m}^{-2}$ (e.g. Koch et al. 2010).

The impact of BC on the Arctic climate was estimated by means of the DRE of atmospheric BC, the BC-in-snow albedo effect, the indirect, and the semi-direct radiative effect. In total they amounted to $-0.20\,\mathrm{W\,m}^{-2}$ at TOA, meaning BC would be cooling in the Arctic. However, the cloud mediated effects are highly uncertain.

The Arctic ($> 60°$ N) TOA DRE of BC, as the main focus of this work, is warming the Arctic. With an annual average of $0.31\,\mathrm{W\,m^{-2}}$ it was slightly higher in this study than the multi-model median of the AeroCom models of $0.19\,\mathrm{W\,m^{-2}}$, but within the range of models (Sand et al. 2017). There, however, only the DRE by the BC increase from pre-industrial levels is given, while here the effect of total BC was considered. For the TOA DRE of BC the uncertainty with regards to both emissions and to an optimisation in the ageing rate and in-cloud scavenging fraction, was estimated at roughly $25\,\%$. This uncertainty appears to be statistically significant at TOA and bottom of the atmosphere (BOA).

The BC-in-snow albedo effect contributed $0.12\,\mathrm{W\,m^{-2}}$ at TOA, which is roughly the same as the previous estimate of Gilgen et al. (2018), even with the adaptations of emissions and long-range transport in this study. The uncertainty with regard to emissions was $15\,\%$ for the BC-in-snow albedo effect, but there was almost no difference in the BC-in-snow albedo effect between the runs with and without the optimisation of wet removal. Both the DRE of atmospheric BC and the BC-in-snow albedo statistically significantly altered the DRE of total aerosol in most of the Arctic.

While the DRE and the BC-in-snow albedo effect affected the Arctic relatively uniformly and caused a net energy gain at TOA, the cloud mediated effects were much more heterogeneous, with regions of net energy gains and losses at TOA. Both, IRE and sDRE were found to be over one order of magnitude larger locally than the Arctic average. The TOA IRE of atmospheric BC was estimated at $-0.30\,\mathrm{W\,m^{-2}}$ and the sDRE at $-0.33\,\mathrm{W\,m^{-2}}$ in the Arctic in this study.

Because clouds have a very strong effect on the net radiation balance, but are very heterogeneously distributed and because of the stochastic nature of their occurrence, neither the two direct radiative effects nor the cloud mediated radiative effects were found to have a statistically significant impact on the net TOA/BOA radiative balance, as also stated by Donth et al. (2020). This does not, however, mean that their impact on the Arctic climate is negligible.

Overall, the main outcomes of this book are:

- Observation-based, daily fire emissions are key for reproducing observed time series and atmospheric profiles, while there is no clear advice on anthropogenic emission datasets.

- Extensive comparisons to airborne measurements proved ECHAM-HAM to be biased high in upper tropospheric BC levels in the Arctic.

- An optimised configuration was found, that enabled a better reproduction of Arctic BC vertical profiles, where the in-cloud scavenging efficiency was doubled and the number of required SU mono-layers for aerosol ageing was increased to two. This leads to a BC lifetime of $5.6\,\mathrm{d}$.

- At TOA, the Arctic average DRE of atmospheric BC in the "BC-Optimised" model run was estimated at $0.31\,\mathrm{W\,m^{-2}}$ annually, having a warming effect on the Arctic, with $0.56\,\mathrm{W\,m^{-2}}$ being absorbed within the atmospheric column on the annual average.

- The uncertainty in emissions was estimated to lead to an uncertainty of $25\,\%$ in the Arctic burden and $22\,\%$ in the TOA DRE of BC.

- The estimate for the uncertainty in removal uncertainties was much lower in the Arctic BC burden at $10\,\%$, but $24\,\%$ in the TOA DRE of BC. This seeming contradiction shows the importance of correctly reproducing the vertical distribution of Arctic BC.

- The effective annual radiative effect of BC, including the warming DRE and BC-in-snow albedo effect as well as the highly uncertain, mainly cooling, cloud-mediated effects (IRE and sDRE), was estimated to be slightly cooling at $-0.2\,\mathrm{W\,m^{-2}}$ in the Arctic.

Further studies on the model uncertainties in Arctic black carbon, which could follow on from this work, could relate to the role of wet convection. Current models have to parametrise convective cells and the related convective cloud formation and precipitation, which is potentially subject to large biases. The recent push to convection-resolving models opens up new possibilities to investigate and improve cloud scavenging processes in global climate models.

Another somewhat related source of uncertainty is the representation of the injection height of biomass burning events. This injection height depends on the strength of the fire, which causes pyroconvection. Strong pyroconvection is capable of lifting aerosol particles far out of the planetary boundary layer. Additionally, there is the possibility of this pyroconvection to cause cloud formation. In some cases stratospheric biomass burning aerosol was observed (Ditas et al. 2018). The GFAS emission data also provides an estimate of release heights, which is not used in the current version of ECHAM-HAM, but might be worth considering in a follow-up study. Here, convection-resolving models could also lead to great insight on how important these events are for the Arctic climate, if they have the capability of representing pyroconvection.

The composition and therefore the optical properties of soot vary with the emission sources. For instance, biomass burning smoke contains different amounts of brown carbon depending on material burned and the temperature-dependent combustion state. The characterisation of the optical properties of mixed fire plumes was beyond the scope of this book but would be desirable to be addressed, e.g., by sensitivity studies and further evaluation.

During the evaluation of how the optimisation for Arctic BC would affect other aerosol concentrations, it was found that organic carbon concentrations were strongly underestimated by ECHAM-HAM compared to ATom aircraft observations. In this regard the ongoing ECHAM-HAM evaluation should also focus on organic carbon. While not as absorbing as BC, it could have an important impact in the aerosol-cloud interactions.

With major biomass burning events becoming more frequent in a continuously warming climate (de Groot et al. 2013), this topic seems to be of great importance. BC from biomass burning might become even more important in the Arctic, if the unprecedented high-latitude wildfire frequency and strength of the last two years becomes a trend (Parrington et al. 2020).